# SpringerBriefs in Applied Sciences and Technology

SpringerBriefs present concise summaries of cutting-edge research and practical applications across a wide spectrum of fields. Featuring compact volumes of 50 to 125 pages, the series covers a range of content from professional to academic.

Typical publications can be:

- A timely report of state-of-the art methods
- An introduction to or a manual for the application of mathematical or computer techniques
- A bridge between new research results, as published in journal articles
- A snapshot of a hot or emerging topic
- An in-depth case study
- A presentation of core concepts that students must understand in order to make independent contributions

SpringerBriefs are characterized by fast, global electronic dissemination, standard publishing contracts, standardized manuscript preparation and formatting guidelines, and expedited production schedules.

On the one hand, **SpringerBriefs in Applied Sciences and Technology** are devoted to the publication of fundamentals and applications within the different classical engineering disciplines as well as in interdisciplinary fields that recently emerged between these areas. On the other hand, as the boundary separating fundamental research and applied technology is more and more dissolving, this series is particularly open to trans-disciplinary topics between fundamental science and engineering.

Indexed by EI-Compendex, SCOPUS and Springerlink.

Jinsheng Wang · Hanin Samara · Philip Jaeger

# Carbon Dioxide Adsorption in Rock and Geological Storage of Carbon

Jinsheng Wang
CanmetENERGY-Ottawa
Natural Resources Canada
Ottawa, ON, Canada

Hanin Samara
Institute of Subsurface Energy Systems
Clausthal University of Technology
Clausthal-Zellerfeld, Germany

Philip Jaeger
Institute of Subsurface Energy Systems
Clausthal University of Technology
Clausthal-Zellerfeld, Germany

ISSN 2191-530X            ISSN 2191-5318   (electronic)
SpringerBriefs in Applied Sciences and Technology
ISBN 978-3-031-90217-8        ISBN 978-3-031-90218-5   (eBook)
https://doi.org/10.1007/978-3-031-90218-5



This Springer imprint is published by the registered company Springer Nature Switzerland AG
The registered company address is: Gewerbestrasse 11, 6330 Cham, Switzerland

If disposing of this product, please recycle the paper.

# Preface

This work is on adsorption of $CO_2$ on rocks, which could play an important role in geological storage of $CO_2$, an essential component of carbon capture and storage (CCS) for achieving zero and negative carbon emissions to the atmosphere.

$CO_2$ is the primary greenhouse gas emitted by industrial and energy-related activities, particularly the burning of fossil fuels. Its accumulation in the atmosphere has contributed to global warming, reaching levels that could become intolerable for human societies and for animal and plant life. Carbon capture aims at reducing $CO_2$ emissions, primarily by capturing it from high-concentration sources like power plants and refineries. It also includes direct air capture (DAC), which extracts $CO_2$ from the atmosphere. While industrial capture addresses emissions at the source, DAC helps reduce $CO_2$ that has already been released into the atmosphere and offset $CO_2$ from sources that are harder to mitigate. For both processes, it is vital to achieve permanent removal of the captured $CO_2$ by storing it in underground porous rock formations such as deep saline aquifers, depleted oil and gas fields, and unmineable coal seams. However, the storage process faces various challenges, including limited storage efficiency, the risk of leakage and cost. A number of measures have been explored to maximize $CO_2$ storage while minimizing the risks of leakage, where understanding and predicting $CO_2$ behavior in the underground formations are essential. The movement of $CO_2$ through porous rocks and its interactions with rock minerals and formation water, which occupy the pores, are not directly observable. Whereas monitoring techniques such as seismic surveys and borehole measurements can provide useful data, they have limitations in spatial resolution and coverage. Predictive models can help simulate the movement and interactions to fill in the gaps and provide insights into the processes. However, an important factor, adsorption of $CO_2$ on pore surface of rocks, has been overlooked. Adsorption in porous solids has wide applications in industry, including adsorbents for $CO_2$ capture. On the other hand, concerning $CO_2$ storage only adsorption in coal and organic-rich shales have been actively investigated. For conventional reservoir rocks sandstones and carbonates, which make a majority of geological formations targeted for $CO_2$ storage, $CO_2$ adsorption has not received due attention. Adsorbed $CO_2$ has higher density and low mobility which can increase $CO_2$ storage capacity and efficiency while decreasing the risk of leakage. It can also decrease the cost of $CO_2$ storage in various aspects including injection, pressure management and monitoring.

However, evaluation of $CO_2$ adsorption on rocks for $CO_2$ storage is difficult. Current methods for adsorption measurements are unable to determine the adsorption capacity properly under $CO_2$ storage conditions. The authors have addressed this issue and developed methods for the evaluation. It is also demonstrated how seemingly limited adsorption capacity of rocks can benefit $CO_2$ storage substantially.

The authors hope that this work will not only raise the awareness of the importance of $CO_2$ adsorption to $CO_2$ storage, but also provide useful tools for research and development for early deployment of CCS, in order to contribute to the Paris Agreement goal and developing a sustainable future.

Ottawa, Canada                                                        Jinsheng Wang
Clausthal-Zellerfeld, Germany                                          Hanin Samara
Clausthal-Zellerfeld, Germany                                          Philip Jaeger

**Acknowledgment** This work is supported by the Program of Energy Research and Development of Canada.

**Competing Interests** The authors have no competing interests to declare that are relevant to the content of this manuscript.

# Contents

# Chapter 1
# Introduction

## 1.1 Overview

The global warming is increasing at an alarming rate. Several days in the summer of 2024 recorded the highest temperature in the history. The cause of the global warming has been attributed to the accumulation of greenhouse gases in the atmosphere. A major contributor to the increased greenhouse gases is carbon dioxide from anthropogenic use of carbon such as burning fossil fuels. Carbon capture and storage (CCS) is one of the most evaluated measures to limit the global warming by reducing the release of carbon dioxide into atmosphere. $CO_2$ storage in geological formations is the key component of CCS to achieve zero and even negative carbon emissions to the atmosphere. Decarbonization of industry may be achieved through replacing fossil fuel energy with renewable energy such as solar and wind energy. However, even if fossil fuels are totally replaced with renewable energy, some important industrial processes such as cement and steel production would continue to generate $CO_2$. Besides, $CO_2$ concentration in the atmosphere is already too high to allow curbing the temperature rise. Large projects for direct air capture, i.e., removing $CO_2$ from atmosphere, are currently funded. This approach needs to be coupled with geological $CO_2$ storage, which provides a sink for the captured $CO_2$. Geological reservoirs are porous rock bodies which could accommodate $CO_2$. The pore space of the reservoirs targeted for $CO_2$ storage is normally occupied by non-potable high-salinity water. The capacity for $CO_2$ storage is proportional to the volume of empty space in porous rocks, which is measured by porosity, the fraction of the empty space in the total volume. Larger porosity enables higher storage capacity. However, to enter the pore space and displace the water, $CO_2$ must be injected under high pressure. The injected $CO_2$ will remain as a fluid phase over very long time due to limited dissolution in reservoir water. Because the density of the $CO_2$ fluid is lower than the density of reservoir water, injected $CO_2$ experiences a buoyancy force and moves upwards until reaching the caprock, an impermeable or low-permeability rock layer. Then the $CO_2$ phase will move laterally along the bottom of the caprock. If the movement of

© The Author(s), under exclusive license to Springer Nature Switzerland AG 2025
J. Wang et al., *Carbon Dioxide Adsorption in Rock and Geological Storage of Carbon*,
SpringerBriefs in Applied Sciences and Technology,
https://doi.org/10.1007/978-3-031-90218-5_1

the $CO_2$ phase is confined by special geological structures such as anticlines, part of the $CO_2$ will be trapped while the un-trapped $CO_2$ will continue spreading. $CO_2$ can dissolve in reservoir water gradually. However, dissolution in reservoirs is a very slow process which may take hundreds of years. Before full dissolution, the $CO_2$ fluid under the overpressure (which refers to any pressure higher than the original reservoir pressure following $CO_2$ injection) and the upward buoyant force can leak from the reservoirs through fractures in caprocks, abandoned wells and other defects (Pawar et al. 2020). Moreover, under high $CO_2$ pressure the caprock can be damaged, increasing the potential for $CO_2$ to leak to the atmosphere or contaminate drinking-water reservoirs (Birkholzer et al. 2012). High injection pressure can also cause seismic events or tremors (Kroll et al. 2020). Accordingly, to ensure the security and safety of $CO_2$ storage, injection of $CO_2$ into geological reservoirs is constrained by allowable overpressure. In order to maintain the pressure within the allowable range while maximizing $CO_2$ storage, careful analyses of the storage mechanisms are crucial.

Adsorption is a well-known phenomenon with numerous industrial applications, including the treatment of sour gases like $CO_2$ and carbon capture. However, it has not been extensively studied in the context of geological $CO_2$ storage. Through adsorption, $CO_2$ can adhere to the pore surfaces of reservoir rocks, forming a dense and immobile layer. This process can increase $CO_2$ storage capacity, reduce $CO_2$ buoyancy, and mitigate overpressure in reservoirs. As a result, more $CO_2$ can be stored, while the risk of leakage is minimized compared to retaining $CO_2$ in its fluid phase. To date, the role of adsorption in $CO_2$ storage geological reservoirs has not been systematically explored, except for research on shale and coal. This book focuses on development and application of evaluation methods based on experimental findings and relevant theories, to assess $CO_2$ adsorption for more secure, efficient, and sustainable storage practices. First, a brief introduction to the current state of $CO_2$ storage will be given, followed by the presentation of key theoretical and experimental approaches, including formulas and models for quantifying adsorption and analyzing how adsorption influences other $CO_2$ storage mechanisms, particularly in enhancing storage efficiency, reducing mobility, and mitigating overpressure risks.

## 1.2  Geological Reservoirs for $CO_2$ Storage

Geological reservoirs considered for $CO_2$ storage include saline aquifers, oil and gas fields, and coal seams. Oil and gas fields have existing infrastructure that can be used for injecting and monitoring $CO_2$, thus reducing costs. Besides, $CO_2$ has already been applied for decades to enhanced oil recovery (EOR) to increase economic profitability. These fields have been evaluated extensively, resulting in good under-standing of their geological properties and potential for long-term $CO_2$ storage. The driving mechanisms taking place in combined EOR-CCS are discussed by Samara et al. (2022) and the technical status is described in ISO 27916:2019(E) that is currently being revised. However, it was found that due to the high solubility of

$CO_2$ in the hydrocarbon phase, an important part of $CO_2$ is produced along with the hydrocarbons. Therefore, the capacity of oil and gas fields is limited and may not be sufficient to accommodate large volumes of $CO_2$. Besides, abandoned wells may provide pathways for $CO_2$ leakage. Coal seams can allow enhanced recovery of methane by injection of $CO_2$. The injected $CO_2$ displaces methane in coal seams, and the methane can then be captured for use while the $CO_2$ is adsorbed by coal and stored. However, coal seams generally have low injectivity. Compared to saline aquifers and oil and gas fields, the capacity of coal seams is much smaller. Saline aquifers have the largest potential for $CO_2$ storage due to appropriate porosity and permeability, potentially allowing for accommodation of large amounts of injected $CO_2$. Deep saline aquifers have large storage capacities and thick and widespread top layers which were formed over millions of years. The weight of overlying materials makes the layers compact and chemical processes like cementation, where minerals precipitate and fill the void spaces, render the layers more impermeable. As a result, the risk of $CO_2$ leakage is low. The disadvantages of saline formations include higher injection costs, monitoring challenges and environmental risks include seismic activities (earthquakes) and contamination of freshwater resources.

Another type of reservoirs are tight reservoirs, also known as unconventional gas or oil reservoirs. They have low permeability and porosity or low pore connectivity, leading to low $CO_2$ injectivity. However, there is still potential for $CO_2$ storage, particularly when combined with enhanced recovery of hydrocarbons. Many tight reservoirs already have infrastructure for $CO_2$ injection and monitoring. The fractures in the reservoirs produced by hydraulic fracturing, a process for increasing the permeability, can increase the injectivity for $CO_2$ and the space for $CO_2$ storage. In shale reservoirs, a major type of tight reservoirs, $CO_2$ can be adsorbed in organic matter of shales. Adsorption of $CO_2$ on clay minerals, which are dominant mineral types in shales, has been studied in the context of the sealing capacity of caprocks (Gasparik et al. 2012).

An important criterion for the selection of reservoirs is their depth. In order to maximize the storage capacity, $CO_2$ needs to be compressed to smaller volumes. The natural pressure in the subsurface increases with depth because of increasing overburden masses of soil, rock and water with the depth. $CO_2$ is in its supercritical state under reservoir conditions, where the temperature is usually well above its critical temperature (31 °C) and the pressure is above the critical pressure of 73.8 bar. Near the critical point, the density of supercritical $CO_2$ increases drastically with increasing pressure. Below the critical pressure and above the critical temperature, $CO_2$ density is much lower, resulting in relatively small quantities of $CO_2$ that can be stored. To enable $CO_2$ to attain a higher density, a minimum depth of 800 m is commonly considered. This is according to the pressure gradient in water which is used to approximate the pressure gradient of the subsurface, as a water column of 800 m high will reach a pressure about 80 bar at the bottom.

Next to the required depth, the volume of the pore space needs to be large to ensure adequate storage capacity. Sedimentary rocks have voids between the consolidated grains which form the pore space and reservoir water fills the space under gravity forces. For $CO_2$ storage to occur, $CO_2$ must be pressurized in order to be forced into the pore space. Injected $CO_2$ and the reservoir water are regarded as immiscible phases, although they can dissolve in each other to a limited degree. At the interface separating the two phases, an interfacial tension force is generated which inhibits the molecules to cross the boundary and disperse in the other phase. If the fluids were fully miscible, the interfacial tension would vanish and the molecules of each species could mix with each other at any ratio as known from miscible flooding of hydrocarbon with compressed $CO_2$ (Ahmadyar and Samara 2025). Due to the limited solubility of $CO_2$ in water, large quantities of water are required to dissolve injected $CO_2$. Besides, large contact area required to allow adequate mass transfer for the dissolution, which is restricted in reservoirs. The interfacial tension between $CO_2$ and water decreases with increasing pressure but maintains a level well above zero, indicating coexistence of two clearly distinguished phases. Across the interface, a pressure difference called capillary pressure could occur which opposes the advance of the $CO_2$ phase in the pores, increasing the difficulty for $CO_2$ to access and occupy more pore space. To overcome the capillary pressure, higher $CO_2$ injection pressure is required, causing greater overpressure.

Another important factor is the injectivity, a measure of the easiness for injecting $CO_2$ into the reservoir. It is characterized by the rate with which $CO_2$ can be injected, and its ability to move away from the injection well (De Silva and Ranjith 2012). Injectivity is expressed as the ratio of the quantity of $CO_2$ that is injected to the difference between the pressure that is required for injecting the respective amount of $CO_2$ and the pressure evolving in the reservoir (Raza et al. 2022a). If the injectivity is low, the quantity of injected $CO_2$ may not be able to reach the capacity of the reservoir, unless very high pressures are applied. High injectivity requires high permeability, a measure of a rock to transmit fluids, as shown by Darcy's law which is commonly used to describe fluid flow in porous rocks:

$$\frac{Q}{A} = k\,\Delta p/\mu L, \tag{1.1}$$

where $k$ is the permeability of rock; $Q$ is the flow rate; $A$ is the cross-sectional area of the rock; $\mu$ is the viscosity of the fluid; $\Delta p$ is the pressure difference over the length $L$. The permeability is determined by porosity, pore size and pore tortuosity according to the Carman-Kozeny law (Graczyk and Matyka 2020). For two-phase flows, the flow rates are determined by the relative permeability, which is the ratio of the effective permeability of the given fluid phase to the intrinsic permeability, i.e., the permeability when one fluid alone flows in the same rock (Islam 2022). Because the two-phase interferes with each other, the relative permeability is lower than the intrinsic permeability. A low relative permeability signifies a low injectivity as well. In that case, either higher injection pressures or more injection wells are required in order to place the required quantity of $CO_2$ in the reservoir.

Still another but more important consideration is the security of the top seal of the reservoir. The targeted reservoir must have a top layer of impermeable or very low-permeability rock, i.e., the caprock, to prevent $CO_2$ leakage. As has been mentioned earlier, injected $CO_2$ is buoyant and moves upward. When stopped by the caprock, the pressure at the top of the CO$_2$ phase will be higher than the pressure in the surrounding water due to the buoyancy. This will result in a greater upthrust on the caprock and further increase the potential for CO$_2$ leakage. Unlike water which tends to sink, CO$_2$ in the reservoir tends to move up wherever possible because of its buoyancy. If pathways for leakage are encountered, CO$_2$ would go through and escape. Accordingly, CO$_2$ caprock integrity must be carefully assessed for storage sites. Besides, pressure for CO$_2$ injection must be carefully controlled to ensure the security. Furthermore, CO$_2$ should be trapped in order to reduce its area of spreading and reduce its chance to encounter leakage pathways.

## 1.3  Mechanisms for Trapping CO$_2$ in Reservoirs

The general purpose of geological CO$_2$ storage is to contain the CO$_2$ in the subsurface and minimize the risks of leakage and environmental damage. CO$_2$ traps can decrease the potential of leakage and contribute to long-term security and safety of CO$_2$ storage. There are four commonly accepted mechanisms for CO$_2$ traps (see, e.g., Bachu 2008; Massarweh and Abushaikha 2024).

### *1.3.1  Structural and Stratigraphic Trapping*

Deeper than 800 m from the surface, injected CO$_2$ would be in the supercritical state with a density from 500 to 800 kg/m$^3$, depending on temperature. Even though the density of supercritical CO$_2$ is high, it is lower than the density of reservoir water or brine which is over 1000 kg/m$^3$. As CO$_2$ dissolution into water is slow, the majority part of injected CO$_2$ will be in a buoyant fluid phase. This fluid phase results in a so-called CO$_2$ plume, which is a region where the CO$_2$ is concentrated and moving within the reservoir. Under the buoyancy effect, CO$_2$ plume rises toward the top of the reservoir and then spreads out until it gets trapped. Structural and stratigraphic trapping occur under impermeable or low-permeability rock layers which could stop upward and lateral movements of CO$_2$. The structural traps are inverted V or dome-shaped regions beneath caprocks of the reservoirs, where CO$_2$ will be confined in the crests. The stratigraphic traps are fluid-confining zones owing to variations in porosity, permeability and lithology within rock layers, which can lead to trapping of CO$_2$. These types of trapping are considered to be the major trapping mechanism in the early stage of CO$_2$ storage.

### 1.3.2 Residual Trapping

Residual trapping refers to trapping $CO_2$ as disconnected droplets and clusters that are immobilized by water due to capillary forces. Sedimentary rocks including sandstones and carbonates which do not contain organic matter are believed to be water-wet naturally. During injection stage, $CO_2$ must displace water in pore space under high pressure. During the later stage after $CO_2$ injection is stopped, water will come back in the wake of $CO_2$ plume, a trail behind the main body where $CO_2$ pressure is lower. The invading water breaks some of the $CO_2$ from the main body and keep the disconnected $CO_2$ in pore space by capillary forces, leading to residual trapping.

This mechanism requires water imbibition (i.e., the process where water displaces non-aqueous fluids) in pore space previously occupied by $CO_2$ following decrease of $CO_2$ pressure. According to a number of analyses, residual trapping could immobilize a significant fraction of injected $CO_2$.

### 1.3.3 Solubility Trapping

Solubility trapping refers to trapping $CO_2$ as dissolved species in the reservoir water. These species include hydrated $CO_2$ molecules, carbonic acid ($H_2CO_3$) and ions including $HCO_3^-$ and $CO_3^{2-}$ according to the following reactions:

$$CO_2(aq) + H_2O \leftrightarrow H_2CO_3(aq) \tag{1.2}$$

$$H_2CO_3 \leftrightarrow H^+ + HCO_3^- \tag{1.3}$$

$$HCO_3^- \leftrightarrow H^+ + CO_3^{2-}. \tag{1.4}$$

$CO_2$ dissolution in water increases with increasing pressure and decreasing temperature (Wiebe and Gaddy 1939). The solution of $CO_2$ in water is weakly acidic with a pH in the range of 3–4 (Almeida et al. 2021). The density of the $CO_2$-saturated aqueous phase is higher than the pure brine (Samara et al. 2022), inducing a flow downwards in the aquifer, whereas free-phase $CO_2$ tends to move upwards, following a complex mechanism of two-phase flow that is affected by the relative permeabilities of the free phase and the liquid phase as well as the capillary pressure (Salimi et al. 2012). Solubility trapping is considered a more secured trapping mechanism than structural trapping, where $CO_2$ remains buoyant and can leak from the top. Dissolved $CO_2$ can react with some rock minerals to form solid carbonate products, resulting in trapping of $CO_2$ in the solids.

Under $CO_2$ storage conditions, i.e., high pressure and moderate temperature (40–80 °C), $CO_2$ dissolution is thermodynamically favored. However, the dissolution requires sufficient contact area between $CO_2$ and water for an efficient mass transfer across the boundary to happen, as mentioned earlier. From the phase boundary onwards, CO$_2$ diffuses along the pores. According to some estimates, dissolution of injected CO$_2$ in the reservoirs can take hundreds of years until saturation conditions are reached.

## *1.3.4  Mineral Trapping*

Mineral trapping refers to trapping CO$_2$ as insoluble solid products. CO$_2$ solutions could react with some aluminum silicate and carbonate rocks to form stable products such as siderite and calcite. These solid carbonate products can precipitate or cover the inner surface of the pore network in reservoir rocks, enabling permanent storage of CO$_2$. Mineral trapping is considered the most secure trapping mechanism for CO$_2$. However, the reactions require dissolved CO$_2$. Accordingly, the rate of the reactions is limited by CO$_2$ dissolution in reservoir water. Thus, mineral trapping will take place even more slowly than solubility trapping.

Following dissolution of CO$_2$, carbonic acid is formed according to the Reaction 1.2. The acid further dissociates, forming H$_3$O$^+$ ions according to Reaction 1.3 that reduce the pH, enhancing dissolution of rock minerals such as anorthite [CaAl$_2$Si$_2$O$_8$], olivine [(Fe,Mg)$_2$SiO$_4$] and feldspar [NaAlSi$_3$O$_8$-CaAl$_2$Si$_2$O$_8$] from the inner rock surface while H$^+$ is bonded and divalent cations such as Ca$^{2+}$ are released forming Kaolinite [Al$_2$Si$_2$O$_5$(OH)$_4$], all in all increasing the pH and converting carbon to carbonate ions (Xiong et al. 2017) that on their turn precipitate as carbonate minerals including ankerite [Ca(Fe,Mg,Mn)(CO$_3$)$_2$], calcite (CaCO$_3$), siderite (FeCO$_3$) and/or magnesite (MgCO$_3$). Mineralization in sandstone formations is less pronounced due to the low content of cations such as Fe$^{3+}$, Mg$^{2+}$ or Ca$^{2+}$ (Kim et al. 2023) while carbonate reservoirs and some shales are rich in Ca$^{2+}$. Basalt contains important amounts of feldspar (Xiong et al. 2017), for which it is seen as an optimal candidate for CO$_2$-mineralization, but here the challenge is the lack of sufficient aqueous phase for dissolving the gas as only in this case carbonic acid can be formed that on its turn is required to dissolve the decisive minerals. However, most research on this topic has been conducted on basalts (Gislason and Oelkers 2014; Raza et al. 2022b) and field tests have been carried out in this type of formation (Aradóttir et al. 2012; Gunnarsson et al. 2018; White et al. 2020). In order to overcome the lack of groundwater in the basalt formation of the Carbfix project, water is injected along with the CO$_2$ (Matter et al. 2009). Next to the lithology, process conditions such as pressure, temperature and pH play a crucial role (Liu and Maroto-Valer 2011; Druckenmiller and Maroto-Valer 2005). Especially the pressure and the pH are related to each other through the solubility of CO$_2$ which is the limiting factor, as the amount of CO$_2$ dissolved in the aqueous phase most likely ranges below 5 wt%, even at elevated pressures (Wiebe and Gaddy 1939). As carbonic acid reacts to form solid species such as calcite, the chain of reactions forces more CO$_2$ to dissolve following the Le Chatelier and Braun principle but within the solubility limit. Hence, the ultimate mineralization step depends on the kinetics of dissolution of CO$_2$, on the presence of divalent ions (e.g., $Ca^{2+}$) as well as the reaction kinetics prior to precipitation. The actual conditions of pressure and temperature not only govern the overall yield through the solubility but take influence on the type of minerals being formed. According to Voigt et al. (2021), under a gas pressure of 16 bar, magnesite was observed to form and precipitate. In comparison, at a pressure of 2.5 bar, calcite was predominantly forming, showing how the combination of pressure and rock composition may lead to different preferential mineralogical deposits.

## 1.4  Operational Issues

### 1.4.1  Pressure Buildup

As the pore space of reservoir rocks are originally filled with reservoir water (and hydrocarbon fluids in the case of depleted oil and gas reservoirs), $CO_2$ must displace the water to enter the pore space. This requires injecting $CO_2$ under high pressure. Because the compressibility of water is low, the pressure in the injection zone increases substantially following injection of $CO_2$. The overpressure propagates faster than the $CO_2$ plume in the reservoir (Bachu 2015), when reservoirs are confined by low-permeability sides, the pore pressure can rise rapidly and transmit to the caprocks. Moreover, the buoyancy force of $CO_2$ will result in additional overpressure. Over long time, injected $CO_2$ will dissolve in brine gradually and the overpressure will dissipate. However, before significant dissolution takes effect, excessive pressure buildup could cause reactivation of faults or upward movement of reservoir water, increasing the risks of earthquakes or contamination of drinking-water reservoirs. In all cases the integrity of $CO_2$ storage will be damaged. To avoid such situations, the injection pressure must be controlled.

### 1.4.2  Low Utilization Efficiency of Reservoir Capacity

Given the huge quantity of $CO_2$ that needs to be stored, sufficient reservoir capacity and injectivity are prerequisite. However, large reservoirs with high injectivity may not be available near large industrial $CO_2$ emitters where $CO_2$ is captured. Transporting $CO_2$ to far-away reservoirs will increase the cost and the risk of $CO_2$ leakage from pipelines. Thus, the capacity of available reservoirs should be fully utilized. However, in reality the capacity utilization is low. As mentioned earlier, the $CO_2$ plume moves upward due to the buoyancy effect, resulting in a large fraction of the lateral pore space uncontacted by $CO_2$. The $CO_2$ plume would take a bell shape in a reservoir (see, e.g., Zhang et al. 2015), where the region under the plume is not touched by $CO_2$.

It should be noted that the $CO_2$ plume corresponds to a volume impacted by $CO_2$, which includes a two-phase zone containing $CO_2$ and reservoir water and is therefore not fully occupied by $CO_2$ (see, e.g., Abdelaal and Zeidouni 2020). Due to the capillary pressure in narrow pores or pore throats, $CO_2$ would bypass the narrow spaces and only go through low-resistance flow channels. Within the plume, the pore space that is actually occupied by $CO_2$ may be under 30% (Causebrook 2012).

In $CO_2$ storage reservoirs, it is desirable for $CO_2$ to occupy more pore space and dissolve completely in water, but when a large fraction of the pore space is not accessed by $CO_2$, the utilization efficiency cannot be high. It has been estimated that even under the most favorable conditions, only about 1–4% of pore space in the reservoir is accessible to $CO_2$ (IEAGHG 2009). Thus, implementation of carbon capture and storage is greatly restricted by the low reservoir utilization.

### 1.4.3  Slow Dissolution of $CO_2$

Ideally one would hope the injected $CO_2$ dissolves instantly into reservoir water, because dissolved $CO_2$ is not buoyant and thus would not rise to the top of the reservoir and leak from there. Moreover, dissolved $CO_2$ increases the volume of the host water to a limited degree, so the pressure increase in the reservoir due to injected $CO_2$ is limited. However, inadequate contact with water in reservoir rocks limits $CO_2$ dissolution. In laboratories, $CO_2$ dissolution is enhanced with mixing tools such as stirrers and under such conditions $CO_2$ is seen to be readily soluble in water. In contrast, in reservoirs the contact between $CO_2$ and water in pore space and dissolution of $CO_2$ would take place at the interfaces between $CO_2$ and water. Because of the lack of mixing means, dissolved $CO_2$ needs to diffuse away from the interfacial layer to leave way for more dissolution. The diffusion coefficient of $CO_2$ in water is in the order of $10^{-9}$ m$^2$/s which is very low. As a result, the diffusion process will be very slow. Moreover, as mentioned earlier, $CO_2$ enters only a small fraction of reservoir pore space and the contact with water is confined in this small fraction. Thus, limited entry of $CO_2$ in rock pores is one of the limiting factors to $CO_2$ dissolution or solubility trapping.

### 1.4.4  Storage Security and Safety

Storage security and safety are of great importance and great public concern. If leakage of $CO_2$ occurs, the integrity of $CO_2$ storage and the efforts on curbing $CO_2$ emissions are undermined. Moreover, $CO_2$ is harmful and could pose health risks and environmental damage.

$CO_2$ leakage may occur through various pathways, including faults and fractures, wellbores, caprocks and permeable layers. Natural faults and fractures in rock formations may act as conduits for leakage if they intersect with storage reservoirs.

Abandoned or improperly sealed wells pose leakage risks when the cement or casing in the wells deteriorates. If the caprock is lack of integrity or fractured due to excessive overpressure, $CO_2$ will leak. If permeable rock layers connect with the storage reservoirs, $CO_2$ may move through them and escape.

Direct exposure to elevated levels of $CO_2$ is dangerous. High concentrations of $CO_2$ can displace oxygen in the air, leading to asphyxiation, which may result in unconsciousness or death. Leaked $CO_2$ may contaminate groundwater, altering its chemistry and affecting drinking-water quality. Increased levels of $CO_2$ can harm plant life and disrupt local ecosystems. Furthermore, incidents of leakage will undermine public confidence in CCS, impeding implement of climate strategies. Ensuring the security and safety of $CO_2$ storage is crucial for the success of CCS initiatives, as it not only protects the environment and public health from potential leakage but also fosters public trust and investment in sustainable technologies to mitigate climate change.

# References


M. Abdelaal, M. Zeidouni, Pressure falloff testing to characterize $CO_2$ plume and dry-out zone during $CO_2$ injection in saline aquifers. Int. J. Greenhouse Gas Control **103**, 10316 (2020). https://doi.org/10.1016/j.ijggc.2020.103160

S. Ahmadyar, H. Samara, The effect of gas and liquid phase composition on miscibility through interfacial tension measurements of model oils in compressed $CO_2$. J. Petrol Explor. Prod. Technol. **15**, 31 (2025). https://doi.org/10.1007/s13202-025-01929-5

A. Almeida da Costa, G. Costa, M. Embiruçu, J.B.P. Soares, J.J. Trivedi, P.S. Rocha, A. Souza, P. Jaeger, The Influence of Rock Composition and pH on Reservoir Wettability for Low-Salinity Water-$CO_2$ Enhanced Oil Recovery Applications in Brazilian Reservoirs, SPE-195982-PA (2021). https://doi.org/10.2118/195982-PA

E. Aradóttir, E. Sonnenthal, G. Björnsson, H. Jónsson, Multidimensional reactive transport modeling of $CO_2$ mineral sequestration in basalts at the Hellisheidi geothermal field, Iceland. Int. J. Greenhouse Gas Control **9**, 24–40 (2012). https://doi.org/10.1016/j.ijggc.2012.02.006

S. Bachu, $CO_2$ storage in geological media: role, means, status and barriers to deployment. Prog. Energy Combus. Sci. **34**(2), 254–273 (2008). https://doi.org/10.1016/j.pecs.2007.10.001

S. Bachu, Review of $CO_2$ storage efficiency in deep saline aquifers. Int. J. Greenhouse Gas Control **40**, 188–202 (2015). https://doi.org/10.1016/j.ijggc.2015.01.007

J.T. Birkholzer, A. Cihan, Q. Zhou, Impact-driven pressure management via targeted brine extraction-conceptual studies of $CO_2$ storage in saline formations. Int. J. Greenhouse Gas Control **7**, 168–180 (2012). https://doi.org/10.1016/j.ijggc.2012.01.001

R. Causebrook, in *An Overview of $CO_2$ Storage Capacity Assessment Methodologies for Saline Aquifers.* IEA CCS Workshop, Sydney, Australia (2012). https://iea.blob.core.windows.net/assets/imports/events/87/Causebrook.pdf. Accessed 10 Feb 2025

M.L. Druckenmiller, M.M. Maroto-Valer, Carbon sequestration using brine of adjusted pH to form mineral carbonates. Fuel Process. Technol. **85**(14–15), 1599–1614 (2005). https://doi.org/10.1016/J.fuproc.2005.01.007

P.N.K. De Silva, P.G. Ranjith, A study of methodologies for $CO_2$ storage capacity estimation of saline aquifers. Fuel **93**, 13–27 (2012). https://doi.org/10.1016/j.fuel.2011.07.004

M. Gasparik, A. Ghanizadeh, P. Bertier, Y. Gensterblum, S. Bouw, B.M. Krooss, High-pressure methane sorption isotherms of black shales from the Netherlands. Energy Fuels **26**, 4995–5004 (2012). https://doi.org/10.1021/ef300405g

S. Gislason, E.H. Oelkers, Carbon storage in basalt. Science **344**(6182), 373–374 (2014). https://doi.org/10.1126/science.1250828

K.M. Graczyk, M. Matyka, Predicting porosity, permeability, and tortuosity of porous media from images by deep learning. Sci. Rep. **10**(1), 21488 (2020). https://doi.org/10.1038/s41598-020-78415-x

I. Gunnarsson, E.S. Aradóttir, E.H. Oelkers, D.E. Clark, M.Þ Arnarson, B. Sigfússon, S.Ó. Snæbjörnsdóttir, J.M. Matter, M. Stute, B.M. Júlíusson, S.R. Gíslason, The rapid and cost-effective capture and subsurface mineral storage of carbon and sulfur at the CarbFix2 site. Int. J. Greenhouse Gas Control **79**, 117–126 (2018). https://doi.org/10.1016/j.ijggc.2018.08.014

IEAGHG, in *Development of Storage Coefficients for Carbon Dioxide Storage in Deep Saline Formations* (2009). https://ieaghg.org/publications/development-of-storage-coefficients-for-carbon-dioxide-storage-in-deep-saline-formations/. Accessed 23 Feb 2025

M.R. Islam, Chapter 7—Field guidelines, in: M.R. Islam (Ed.), *Reservoir Development, Sustainable Oil and Gas Development Series*. Gulf Professional Publishing, pp. 737–843 (2022). https://doi.org/10.1016/B978-0-12-820053-7.00004-4

K. Kim, D. Kim, Y. Na, Y. Song, J. Wang, A review of carbon mineralization mechanism during geological $CO_2$ storage. Heliyon **9**(12), e23135 (2023). https://doi.org/10.1016/j.heliyon.2023.e23135

K.A. Kroll, T.A. Buscheck, J.A. White, K.B. Richards-Dinger, Testing the efficacy of active pressure management as a tool to mitigate induced seismicity. Int. J. Greenhouse Gas Control **94**, 102894 (2020). https://doi.org/10.1016/j.ijggc.2019.102894

Q. Liu, M.M. Maroto-valer, Parameters affecting mineral trapping of $CO_2$ sequestration in brines. Greenhouse Gases Sci. Technol. **1**(3), 211–222 (2011). https://doi.org/10.1002/ghg.29

J.M. Matter, W. Broecker, M. Stute, S. Gislason, E. Oelkers, A. Stefánsson et al., Permanent carbon dioxide storage into basalt: the CarbFix pilot project, Iceland. Energy Proc. **1**(1), 3641–3646 (2009). https://doi.org/10.1016/j.egypro.2009.02.160

O. Massarweh, A. Abushaikha, $CO_2$ sequestration in subsurface geological formations: a review of trapping mechanisms and monitoring techniques. Earth-Science Rev **253**, 104793 (2024). https://doi.org/10.1016/j.earscirev.2024.104793

R.J. Pawar, S. Chu, N. Makedonska, T. Onishi, D. Harp, Assessment of relationship between post-injection plume migration and leakage risks at geologic $CO_2$ storage sites. Int. J. Greenhouse Gas Control **101**, 103138 (2020). https://doi.org/10.1016/j.ijggc.2020.103138

A. Raza, M. Arif, G. Glatz, M. Mahmoud, M. Al Kobaisi, S. Alafnan, S. Iglauer, A holistic overview of underground hydrogen storage: Influencing factors, current understanding, and outlook. Fuel **330**, 125636 (2022a). https://doi.org/10.1016/j.fuel.2022.125636

A. Raza, G. Glatz, R. Gholami, M. Mahmoud, S. Alafnan, Carbon mineralization and geological storage of $CO_2$ in basalt: Mechanisms and technical challenges. Earth Sci. Rev. **229**, 104036 (2022b). https://doi.org/10.1016/j.earscirev.2022.104036

H. Salimi, K.-H. Wolf, J. Bruining, The influence of capillary pressure on the phase equilibrium of the $CO_2$–water system: application to carbon sequestration combined with geothermal energy. Int. J. Greenhouse Gas Control **11S**, S47–S66 (2012). https://doi.org/10.1016/j.ijggc.2012.09.015

H. Samara, M. Al-Eryani, P. Jaeger, The role of supercritical carbon dioxide in modifying the phase and interfacial properties of multiphase systems relevant to combined EOR-CCS. Fuel **323**, 124271 (2022). https://doi.org/10.1016/j.fuel.2022.124271

M. Voigt, C. Marieni, A. Baldermann, I.M. Galeczka, D. Wolff-Boenisch, E.H. Oelkers, S.R. Gislason, An experimental study of basalt–seawater–$CO_2$ interaction at 130 °C. Geochim. Cosmochim. Acta **308**, 21–41 (2021). https://doi.org/10.1016/j.gca.2021.05.056

S.K. White, F.A. Spane, H.T. Schaed, Q.R. Miller, M.D. White, J. Horner, B.P. McGrail, Quantification of $CO_2$ mineralization at the Wallula Basalt Pilot project. Environ. Sci. Technol. **54**(22), 14609–14616 (2020). https://doi.org/10.1021/acs.est.0c05142

R. Wiebe, V.L. Gaddy, The solubility in water of carbon dioxide at 50, 75 and 100 °C, at pressures to 700 atmospheres. J. Am. Chem. Soc. **61**, 315–318 (1939). https://doi.org/10.1021/ja01871a025

W. Xiong, R.K. Wells, A.H. Menefee, P. Skemer, B.R. Ellis, D.E. Giammar, $CO_2$ mineral trapping in fractured basalt. Int. J. Greenhouse Gas Control **66**, 204–217 (2017). https://doi.org/10.1016/j.ijggc.2017.10.003

Z.F. Zhang, S.K. White, M.D. White, Delineating the horizontal plume extent and $CO_2$ distribution at geologic sequestration sites. Int. J. Greenhouse Gas Control **43**, 141–148 (2015). https://doi.org/10.1016/j.ijggc.2015.10.018

# Chapter 2
# Adsorption on Solids and the Potential of CO$_2$ Adsorption in Reservoir Rocks

## 2.1 The Phenomenon of Adsorption

Adsorption is defined as the accumulation of molecules (adsorbate) on the surface of a solid (adsorbent) by means of intermolecular interactions, resulting in the formation of a layer that possesses a higher density than that of the bulk phase surrounding the solid (Rani et al. 2019). Adsorption is caused by electrostatic interactions between the surface and the fluid molecules. There are physical adsorptions and chemical adsorptions. Chemical adsorptions are similar to chemical reactions where chemical bonds are formed. In physical adsorptions, the interactions between the surface and adsorbed molecules are weak and no chemical bonds are involved. Within physical adsorptions, adsorption of nonpolar molecules such as $CO_2$ and $N_2$ is weaker than adsorption of polar molecules such as $SO_2$ and water. The forces that cause physical adsorption are van der Waals forces which are a type of weak electrostatic forces and result in attraction and repulsion between molecules. London dispersion forces, a type of van der Waals forces, are attractive forces resulting from temporary dipoles in nonpolar molecules. Electrons in the molecules are constantly moving, creating temporary dipoles. Away from the molecules, the effects of dipoles cancel out, but when the molecules get close enough, attractions between dipoles become significant. However, when the molecules get too close, repulsion occurs. This also applies to interactions between gas molecules and the solid surface.

London forces are short ranged. The magnitude of the forces is sensitive to the proximity of the adsorbate (adsorbed) molecules from the adsorbent (adsorbing material).

## 2.2 Adsorption Isotherm Patterns and Models

Physical adsorption can result in different patterns of isotherm, which is the relationship between the amount of adsorbate adsorbed per unit mass of adsorbent and the pressure of the adsorbate at a constant temperature. Six types of adsorption isotherms are defined by the International Union of Pure and Applied Chemistry (IUPAC), which are shown in Fig. 2.1.

Type I is a typical adsorption isotherm for porous materials. The curve rises sharply at low pressures, then levels off at high pressures. Type II isotherm shows an initial rise followed by a slower increase in adsorption with pressure, but then the rate of adsorption increases with increasing pressure. Type III isotherm exhibits a weak initial adsorption, but the rate increases steeply with increasing pressure. It does not level off at high pressures. Type IV shows a hysteresis in desorption (i.e., the adsorbed substance is released when the pressure decreases) where the isotherm does not follow the same path of the adsorption. Type V is similar to Type IV with a hysteresis in desorption, but starts with increasing rate of adsorption and then plateaus at high pressures. Type VI isotherm rises stepwise, which is related to discrete layers of adsorption.

One of the most used adsorption models is the Langmuir model (see, e.g., Ruthven 1984), which describes type I isotherms. The model is based on several assumptions:

1. Adsorbed molecules only form a single layer on the solid surface.

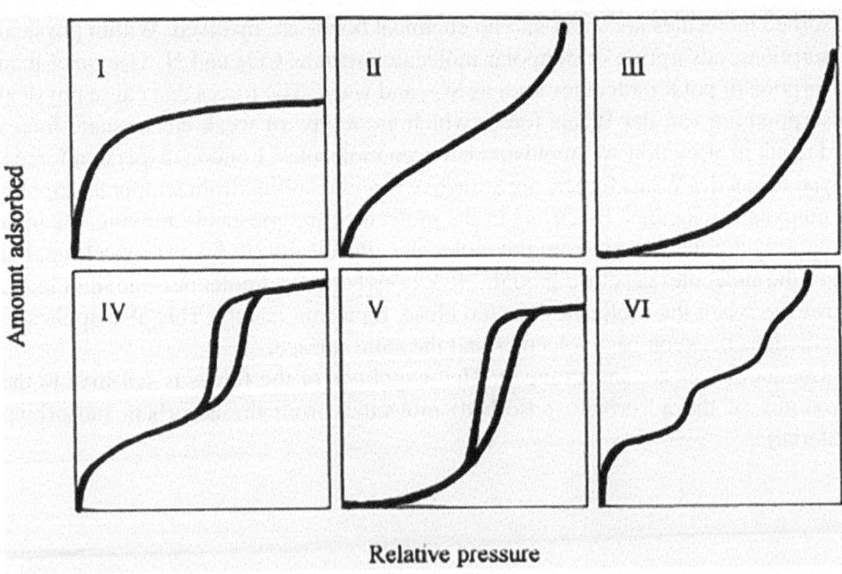


**Fig. 2.1** IUPAC classification of adsorption isotherms for gas–solid equilibria (from Donohue and Aranovich 1998)

2.  The number of adsorption sites on the surface is fixed. When all the sites are taken, no more adsorptions will occur.
3.  The adsorbed molecules do not interact with each other.
4.  Adsorption of gas molecules onto the surface and desorption from the surface back to the gas phase occur simultaneously.

With the above assumptions, the Langmuir model expresses the equilibrium state of the adsorption and desorption as

$$k_a p(1 - \theta) = k_d \theta, \tag{2.1}$$

where the symbol $\theta$ denotes the fraction of adsorption sites occupied by adsorbed molecules. $k_a$ and $k_d$ are rate constants for adsorption and desorption, respectively. $p$ is gas pressure. The left-hand side of this equation corresponds to the rate of adsorption, which is proportional to the pressure $p$ and the fraction of the vacant sites. The right-hand side corresponds to the rate of desorption, which is proportional to the fraction of the sites that have been taken by adsorbed molecules. Rearranging Eq. (2.1) into

$$\theta = \frac{k_a p}{k_a p + k_d} \tag{2.2}$$

leads to

$$C = \frac{C_M b p}{1 + b p}, \tag{2.3}$$

where $C$ represents the amount of adsorbed gas per unit weight or volume of the adsorbent; $C_M$ is the Langmuir capacity, i.e., the maximum amount of adsorbate that could be adsorbed per unit weight or volume of the adsorbent; $C/C_M = \theta$. $b\ (= k_a/k_d)$ in Eq. (2.3) is called the Langmuir adsorption constant, which is dependent on temperature but not on pressure. Whereas the Langmuir model is widely used to describe adsorption data and analyze adsorption behavior, it will be shown that it cannot be applied to $CO_2$ adsorption results in the present work.

Another theory for adsorption is the Polanyi adsorption theory. According to Polanyi, the adsorbed molecules are attracted to solid surfaces as a result of the van der Waals forces near the surfaces (Polanyi 1963; Butt et al. 2006). The attraction forces compress the molecules until condensation occurs and the molecules form a liquid-like layer. The strength of the attraction forces is reflected by an adsorption potential, which can be calculated from the ratio of the equilibrium pressure in the bulk gas phase to the condensation pressure. The potential is given by

$$U(x_0) = -RT \ln\left(\frac{p_0}{p}\right), \tag{2.4}$$

where $U$ is the potential energy for adsorption which is a function of distance from the solid surface. $x_0$ is the thickness of the adsorbed layer. $p_0$ is the condensation pressure. $p$ is the bulk gas pressure. Whereas the Polanyi adsorption theory was originally developed for gases which could condense at the saturation pressure, it has been extended to gases above their critical temperature where saturation pressure is not defined. Although above the critical temperature gases do not condense, the interactions at the solid–gas interface remain similar and the fundamental forces driving the adsorption still apply.

A major difference between Langmuir's model and Polanyi's model is that the former assumes a monolayer of adsorbed phase and the latter considers an adsorbed layer whose thickness can grow in accordance with the adsorption potential. It will be shown that Polanyi's theory reflects better the situation of $CO_2$ adsorption in porous rocks, and this theory will be used hereafter in evaluations for the effects of the adsorption.

## 2.3  The Importance of Adsorption to $CO_2$ Storage

All the $CO_2$-trapping mechanisms mentioned earlier take place in pore space of reservoir rocks. $CO_2$ storage capacity in terms of pore space can be expressed as

$$M = \rho V (1 - S_w), \tag{2.5}$$

where $M$ is the mass of $CO_2$, $\rho$ is the density of $CO_2$, and $V$ is the pore volume. $S_w$ denotes water saturation, which is the fraction of the pore volume occupied by water. The sum of the water saturation and $CO_2$ saturation is equal to 1 if no other fluids are present. Consequently, the term $(1 - S_w)$ represents the fraction of the pore volume that is available for $CO_2$. As can be understood, to increase $CO_2$ occupancy in the pore space, increasing $CO_2$ density and decreasing water saturation are desirable. First, higher density enables more $CO_2$ to be accommodated in given pore volume. The density of $CO_2$ increases with increasing pressure and increases drastically around 80 bar, as shown in Fig. 2.2. This is the reason that geological formations should be over 800 m deep for $CO_2$ storage. In shallower formations $CO_2$ cannot be kept in a dense phase. On the other hand, when the pressure reaches above 10 to 15 MPa, the pressure dependence of $CO_2$ density is increasingly weaker. Conversely, the pressure increases rapidly with increasing density. This behavior puts a limit for $CO_2$ storage: Higher density will increase $CO_2$ occupancy in pore space, but it requires much higher pressure, which can impair storage integrity. Lower temperature will result in higher density, but below 800 m from the surface the temperature cannot be very low due to the geothermal gradient, which is usually 20–30 °C per kilometer depth. Therefore, the temperature increases with depth in the subsurface, which counters the increase of $CO_2$ density.

The other factor in Eq. (2.5), water saturation is controlled by wettability of rocks. Reservoir rocks are usually water-wet, where water adheres to the pore surface and results in capillary forces which oppose the entry of $CO_2$ into the pores. To enable $CO_2$ to enter the pores, higher pressure is required. The pressure increases with decreasing

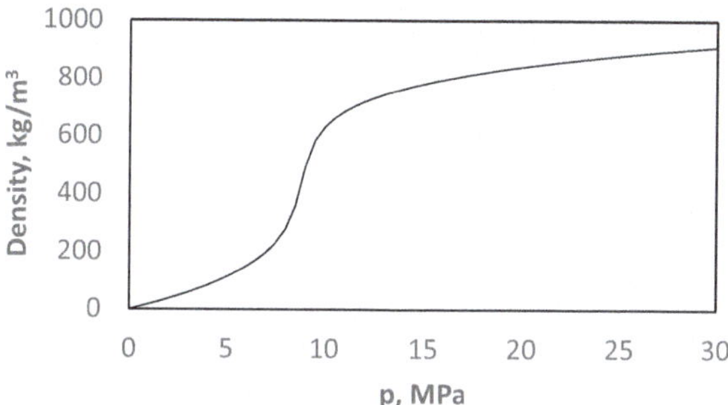

**Fig. 2.2** CO$_2$ density at 40 °C as a function of pressure

pore size, making decreasing water saturation increasingly difficult, as suggested by the capillary pressure curves in Fig. 2.3. Moreover, there is an irreducible water saturation level, beyond which reduction of water saturation is not achievable no matter how high the applied pressure will be. Thus, the capacity of CO$_2$ storage is also limited by water saturation, unless the wettability changes.

If CO$_2$ can be adsorbed on the pore walls, it will form a denser phase because adsorbed molecules are packed more densely. Formation of a denser phase will also decrease the pressure of the free phase because the vacated space allows the free-phase molecules to move in and increase their volume. Adsorption can also change the wettability and increase CO$_2$ saturation in rocks as will be discussed later. The question is whether the pore walls can adsorb a significant amount of CO$_2$ to influence CO$_2$ storage.

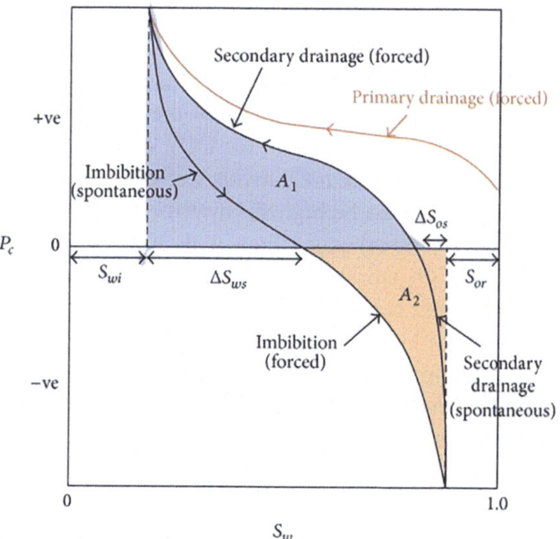


**Fig. 2.3** Illustration of the effect of capillary pressure on water saturation (from Falode and Manuel 2014)

## 2.4 Previous Studies on $CO_2$ Adsorption in Rocks

Adsorption of gases by porous materials has wide industrial applications such as gas separation, gas purification and pollutant removal. Porous materials can also adsorb liquid constituents and are applied in water treatment and other fields. Adsorption of $CO_2$ has been investigated for $CO_2$ capture (Soo et al. 2024). For subsurface applications, adsorption of $CO_2$ in coals and shales has been widely investigated for enhanced productions of coalbed methane and shale gas and oil with $CO_2$ storage. It is well reported that $CO_2$ can be adsorbed by coals and organic matter in shales (see, e.g., Intergovernmental Panel on Climate Change (IPCC) 2018). However, regarding $CO_2$ adsorption in sandstone and carbonate rocks which constitute a majority part of reservoir rocks, very limited studies exist. Coals will not be discussed further as they are predominantly organic matter where the adsorption behavior and effects can differ greatly from those of common reservoir rocks.

### 2.4.1 $CO_2$ Adsorption on Tight Rocks

Tight reservoirs have low permeability and hence low injectivity. However, tight gas and oil reservoirs with horizontal wells and hydraulic fractures could allow $CO_2$ injection. Shales are a major source of tight reservoirs, although tight sandstones and tight carbonates can also make reservoir rocks. While shales are known to be able to adsorb $CO_2$ on the organic matter, inorganic matter like clays can also adsorb $CO_2$ (Busch et al. 2008). Tight rocks are generally associated with small pore size, which can lead to multilayer adsorption and high pore occupancy of adsorbed molecules. In tight reservoirs, the limiting factor for $CO_2$ movement is mass transport. Once $CO_2$ has penetrated into matrixes, the adsorption can be fast (Shen et al. 2024). In tight reservoirs, the thickness of the hydrocarbon producing layers is generally limited, giving limited room for $CO_2$ storage. However, because of the associated short vertical distance the buoyancy effect will also be limited. As a result, the reservoir utilization can be higher. Nevertheless, risks of $CO_2$ leakage (such as through hydraulic fractures) exist (Shen et al. 2024). Adsorption will be the principal mechanism for $CO_2$ storage, and the higher density of adsorbed $CO_2$ phase will allow higher storage capacity. $CO_2$ storage in tight oil and gas reservoirs has economic incentives as hydrocarbon recovery rates in these reservoirs are low and injection of $CO_2$ can produce more saleable hydrocarbons. Combining the enhancement of hydrocarbon recovery with $CO_2$ storage may generate revenues to offset the costs of $CO_2$ storage.

In shale gas reservoirs, a fraction of methane gas exists in an adsorbed state on the organic matter. It has been shown that $CO_2$ has higher affinity than methane to the organic matter. As a result, more $CO_2$ can be adsorbed than methane that is produced. This could enable high storage efficiency of $CO_2$ and offset the greenhouse gas footprint of shale gas, as methane is a potent greenhouse gas (Wang et al. 2011;

Liu et al. 2021). In the case of shale oil, $CO_2$ can displace the oil (Liu et al. 2017; Wang et al. 2020). It was observed in experiments for $CO_2$ adsorption in shale that the mass of the shale sample decreased under high-pressure $CO_2$ and oil came out of the sample, evidencing the ability of $CO_2$ to extract oil from shale and create space for $CO_2$ storage (Wang et al. 2020).

### *2.4.2  $CO_2$ Adsorption on Conventional Reservoir Rocks*

Whereas conventional reservoir rocks sandstones and carbonates do not contain organic matter, they can adsorb $CO_2$ like some other inorganic materials do. Several natural inorganic materials are used as adsorbents for $CO_2$ owing to their porous structures and chemical properties. These materials include natural zeolites such as clinoptilolite and mordenite, natural clays such as bentonite and kaolinite, magnesium silicates such as olivine and serpentinite, and also calcite. Porous reservoir rocks have large pore surface area, where $CO_2$ could be adsorbed as result of unsaturated force fields near the surface. However, there are only a handful of studies on $CO_2$ adsorption in conventional reservoir rocks compared to shales (Wang et al. 2022). In addition to the fact that adsorption in shales has higher economic incentives because of the potential of enhanced oil and gas recovery, apparently low adsorption capacity of conventional rocks could be a reason for the limited studies. Tajnik et al. (2013) studied $CO_2$ adsorption at 21 °C (below the critical temperature of $CO_2$ 31 °C) on several rocks including sandstone and carbonate and found that the adsorption capacities were lower compared to coal. Fujii et al. (2009a, b) studied $CO_2$ adsorption on a sandstone and a granite from 33 °C to 50 °C and reported moderate adsorption capacities. Other published studies on non-shale rocks have primarily focused on enhanced hydrocarbon recovery and caprock sealing capacity.

Experimental results by Wang et al. (2022) for $CO_2$ adsorption in two non-shale reservoir rocks show that the adsorption level ranged from 1 to 3 wt% of the rock samples. While these values are lower than those observed in a shale (which exceeds 4 wt%, as shown in Fig. 2.4), the adsorption capacity of these non-shale rocks is significant for $CO_2$ storage. This is because, under high-pressure conditions, the observed adsorption is lower than the true adsorption capacity. The deviation is due to the higher density of the adsorbed $CO_2$ layer, which cannot be directly measured and results in lower apparent adsorption values. This effect is particularly pronounced for $CO_2$.

Analyses of the results reveal that the actual adsorption level of $CO_2$ can be much higher and the impact of the adsorption on $CO_2$ storage can be substantial. In mainstream assessments for $CO_2$ storage capacity in conventional reservoirs, only pore space or pore volume is considered and the role of pore surface is not taken into account. Some effects of pore surface, such as wettability and chemical reactions, are discussed for $CO_2$ residual trapping and injectivity but no attention is given to adsorption. It will be shown that adsorption of $CO_2$ on rock pore surface reduces $CO_2$ pressure and mobility, which will benefit $CO_2$ storage. The adsorbed $CO_2$ adheres

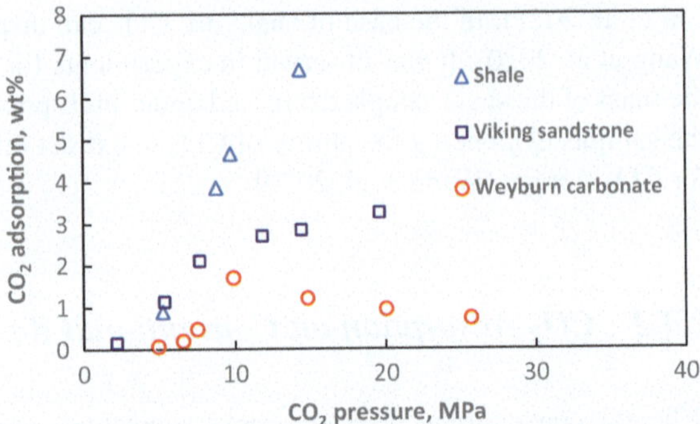

**Fig. 2.4** $CO_2$ adsorption in Canadian reservoir rocks as a function of pressure (data from Wang et al. 2022)



to the pore walls and does no move upward with the free-phase $CO_2$. As a result, a portion of the injected $CO_2$ is immobilized and the mobile $CO_2$ phase is reduced. This will increase reservoir utilization and decrease the potential of $CO_2$ leakage through the top seal. We will briefly introduce the adsorption behavior and measurement methods to explain why the observed apparent adsorption is lower than the actual adsorption under high pressures and then show that $CO_2$ adsorption in reservoir rocks could influence profoundly all operational issues of $CO_2$ storage mentioned earlier. Moreover, we will show that interactions of pore surface with $CO_2$ and water could be largely attributed to $CO_2$ adsorption and relate adsorption to rock wettability, a crucial factor for $CO_2$ storage.

# References


A. Busch, S. Alles, Y. Gensterblum, D. Prinz, D.N. Dewhurst, M.D. Raven, H. Stanjek, B.M. Krooss, Carbon dioxide storage potential of shales. Int. J. Greenh. Gas Control. **2**(3), 297–308 (2008). https://doi.org/10.1016/j.ijggc.2008.03.003

H.J. Butt, K. Graf, M. Kappl, *Physics and Chemistry of Interfaces* (Wiley, New York, 2006)

M.D. Donohue, G.L. Aranovich, Classification of Gibbs adsorption isotherms. Adv. Coll. Interface. Sci. **76–77**, 137–152 (1998). https://doi.org/10.1016/S0001-8686(98)00044

O. Falode, E. Manuel, Wettability effects on capillary pressure, relative permeability, and irreducible saturation using porous plate. J. Pet. Eng. (2014). https://doi.org/10.1155/2014/465418

T. Fujii, Y. Sato, H. Lin, H. Inomata, T. Hashida, Evaluation of $CO_2$ sorption capacity of rocks using a gravimetric method for $CO_2$ geological sequestration. Energy Proc. **1**(1), 3723–3730 (2009). https://doi.org/10.1016/j.egypro.2009.02.171

T. Fujii, Y. Sugai, K. Sasaki, T. Hashida, Measurements of $CO_2$ sorption on rocks using a volumetric technique for $CO_2$ geological storage. Energy Proc. **1**(1), 3715–3722 (2009). https://doi.org/10.1016/j.egypro.2009.02.170

Intergovernmental Panel on Climate Change (IPCC), in *Underground Geological Storage*. IPCC Special Report on Carbon dioxide Capture and Storage. https://www.ipcc.ch/site/assets/uploads/2018/03/srccs_chapter5-1.pdf

B. Liu, C. Wang, J. Zhang, S. Xiao, Z. Zhang, Y. Shen, B. Sun, J. He, Displacement mechanism of oil in shale inorganic nanopores by supercritical carbon dioxide from molecular dynamics simulations. Energy Fuels **31**(1), 738–746 (2017). https://doi.org/10.1021/acs.energyfuels.6b02377

D. Liu, Y. Li, S. Yang, R.K. Agarwal, $CO_2$ sequestration with enhanced shale gas recovery. Energy Sources Part A Recov. Util. Environ. Effects **43**(24), 3227–3237 (2021). https://doi.org/10.1080/15567036.2019.1587069

M. Polanyi, The potential theory of adsorption: authority in science has its uses and its dangers. Science **141**(3585), 1010–1013 (1963). https://doi.org/10.1126/science.141.3585.1010

S. Rani, E. Padmanabhan, B.K. Prusty, Review of gas adsorption in shales for enhanced methane recovery and $CO_2$ storage. J. Petrol. Sci. Eng. **175**, 634–643 (2019). https://doi.org/10.1016/j.petrol.2018.12.081

D.M. Ruthven, *Principles of Adsorption and Adsorption Processes* (Wiley, New York, NY, 1984)

W. Shen, T. Ma, L. Zuo, X. Yang, J. Cai, Advances and prospects of supercritical $CO_2$ for shale gas extraction and geological sequestration in gas shale reservoirs. Energy Fuels **38**(2), 789–805 (2024). https://doi.org/10.1021/acs.energyfuels.3c03843

X.Y.D. Soo, J.J.C. Lee, W.-Y. Wu, L. Tao, C. Wang, Q. Zhu, J. Bu, Advancements in $CO_2$ capture by absorption and adsorption: a comprehensive review. J. $CO_2$ Util. 81, 102727 (2024). https://doi.org/10.1016/j.jcou.2024.102727

T. Tajnik, L.K. Bogataj, E. Jurač, C. Lasnik, J. Likar, B. Debelak, Investigation of adsorption properties of geological materials for $CO_2$ storage. Int. J. Energy Res. **37**(8), 952–958 (2013). https://doi.org/10.1002/er.2901

J. Wang, D. Ryan, E.J. Anthony, Reducing the greenhouse gas footprint of shale gas. Energy Policy **39**(12), 8196–8199 (2011). https://doi.org/10.1016/j.enpol.2011.10.013

J. Wang, D. Ryan, H. Samara, P. Jaeger, Interactions of $CO_2$ with hydrocarbon liquid observed from adsorption of $CO_2$ in organic-rich shale. Energy Fuels **34**(11), 14476–14482 (2020). https://doi.org/10.1021/acs.energyfuels.0c02774

J. Wang, H. Samara, P. Jaeger, V. Ko, D. Rodgers, D. Ryan, Investigation for $CO_2$ adsorption and wettability of reservoir rocks. Energy Fuels **36**(3), 1626–1634 (2022). https://doi.org/10.1021/acs.energyfuels.0c02774

# Chapter 3
# Adsorption Measurements Under High Pressure

Two commonly used methods for adsorption measurement are the gravimetric method and the volumetric method. The gravimetric method measures the mass gain of adsorbent (the material that adsorbs other substances) due to adsorption, usually by use of a high-precision balance. Volumetric method measures the pressure change of gases in a closed system consisting of a reference cell and a sample cell, where the gases are in contact with an adsorbent under pressure. These methods have been long established and well documented. The results of adsorption can be expressed as weight percentage or volume percentage of the adsorbed species over the solid sample.

The same principles are used for adsorption measurements under high pressure. In this chapter, the most commonly applied gas adsorption measurement techniques are covered. These techniques are either gravimetric or volumetric. Below, the respective experimental setups are discussed as well as the operation principle and procedure associated with each method. The discussion is limited to measurements pertaining to pure gases. For gas mixtures, the methods are described elsewhere (Dreisbach and Lösch 2000; Keller et al. 1999). Commercially available high-pressure adsorption measurement equipment is listed in Table 3.1.

## 3.1 The Gravimetric Method

### 3.1.1 Experimental Setup

The gravimetric method relies on the use of a high-resolution balance for the determination of the gas uptake. Although variations of the microbalance exist, e.g., beam balance (Minnick et al. 2018) and magnetic suspension balance (MSB) (Dreisbach and Lösch 2000), the discussion here is limited to the latter configuration, given its

© The Author(s), under exclusive license to Springer Nature Switzerland AG 2025

J. Wang et al., *Carbon Dioxide Adsorption in Rock and Geological Storage of Carbon*,
SpringerBriefs in Applied Sciences and Technology,
https://doi.org/10.1007/978-3-031-90218-5_3

**Table 3.1** Commercially available sorption measurement equipment

| Supplier | Model | Technique | Production origin |
| --- | --- | --- | --- |
| Anton Paar | iSorb HP | Volumetric | Austria |
| TA instruments | isoSORP | Gravimetric | Germany |
| Eurotechnica | HP adsorption unit | Volumetric | Germany |
| Setaram | GASPRO | Volumetric | France |
| MicrotracBEL | BELSORP HP | Volumetric | Japan |
| BSD instrument | BSD-PH | Volumetric | China |
| Hiden Isochema | Xemis, IMI and IGA | Volumetric + Gravimetric | United Kingdom |
| RubolAB | RuboSORP | Gravimetric | Germany |

popularity. The entire setup comprises the MSB, a vacuum pump used to evacuate the system and a gas dosing device, normally a syringe pump, connected to the gas supply to control the gas pressure. The heating of the MSB can be either achieved using an electrical heating jacket or a double-walled heating jacket connected to a thermal bath, depending on the temperature range of interest. A schematic diagram of the system is presented in Fig. 3.1. A computer serves to periodically record the measured mass using a designated software, which can in turn also log experimental conditions transmitted by highly accurate pressure and temperature sensors.

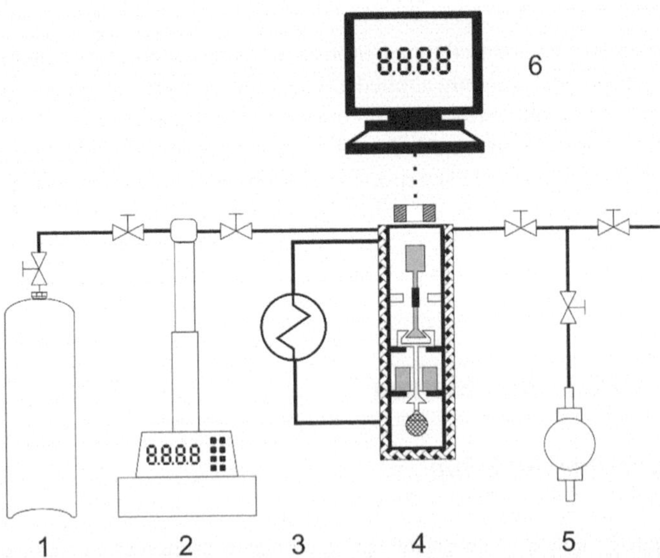

**Fig. 3.1** Adsorption measurement setup based on the gravimetric technique. (1) Gas bottle. (2) Syringe pump. (3) Circulating heating bath. (4) Magnetic suspension balance. (5) Vacuum pump. (6) Computer

### *3.1.2  The Magnetic Suspension Balance*

The MSB comprises mainly two major components that are physically separated: an external part that is subject to atmospheric conditions, and a measurement cell (also referred to as sorption chamber), where experimental conditions are applied. The external part comprises an electromagnet hung from the underfloor weighing hook of a microbalance. The measuring cell comprises an upper permanent magnet connected to an upper suspension shaft, a position sensor, a load coupling system, a lower suspension shaft and a sample crucible. The measuring cell can also be supplemented with a sinker that is attached to the lower suspension shaft. The sinker possesses a precisely known volume and is made of an inert material, either titanium or quartz, possessing low surface to volume ratio (Yang et al. 2023). The sinker serves to locally measure the density of the gas at the applied experimental conditions based on Archimedes' principle. The components of the MSB are shown in Fig. 3.2. The transmission of forces occurs in a contactless manner between the electromagnet (outside) and the permanent magnet (inside) through the walls of the measuring cell (Kleinrahm et al. 2019). The permanent magnet in turn is connected to the sample crucible through the coupling system as shown in Fig. 3.2. The position sensor serves to maintain the permanent magnet in a certain position by signaling any change to a PID controller which modulates the voltage on the electromagnet. Accordingly, the weight is transmitted to the balance via the electromagnet and the mass uptake can be measured (Weireld et al. 1999).

There are principally two positions of the permanent magnet in suspension: the zero-point (ZP) position and the measuring point ($MP_1$) position. These positions change periodically during the measurement by means of the control unit and each of these positions serves a certain purpose. At zero-point (ZP) position (Fig. 3.2a), the permanent magnet shaft is in a low and free suspension state without contacting the load coupling system. The latter remains resting on a support. At ZP, the measured mass is the mass of the permanent magnet suspension (Dreisbach and Lösch 2000), which is calibrated and tared at the beginning of the experiment. During the experiment, the change in zero point is periodically measured enabling the elimination of the balance drift and therefore gaining experimental results of high accuracy. At $MP_1$, the permanent magnet levitates at a higher vertical position (Fig. 3.2b) such that it is in contact with the load coupling system and consequently the sample crucible. Accordingly, the mass of the sample, sample crucible and load coupling system can be measured. In MSBs supplemented with a second measurement point ($MP_2$), the entire system is lifted to a higher position, lifting along a sinker that rests on a support during $MP_1$ measurement. Accordingly, the measured mass at $MP_2$ includes the masses of all the components; what is previously measured at $MP_1$ in addition to the mass of the sinker (Fig. 3.2c).

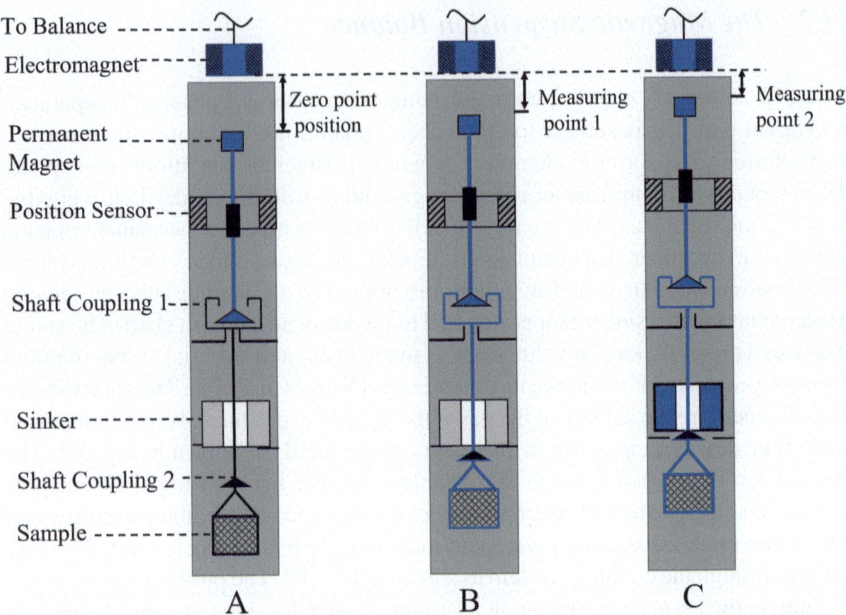



**Fig. 3.2** Internal components of the magnetic suspension balance. Figure is adapted from and modified after TA Instruments brochure for isoSORP series, used with permission. **a** Zero-point position. **b** Measuring point 1 position and **c** measuring point 2 position

### 3.1.3   Experimental Procedure

The experimental procedure is normally conducted in four consecutive steps. The first step comprises a measurement performed at vacuum and at several pressures using helium (He) gas, without the presence of a sample inside the sample crucible. This step is often termed the blank run. The aim of this step is to determine the mass and volume of system components which are required for the calculation of the adsorbed amount, considering the influence of buoyancy on the system. While the determination of the mass is straightforward, the determination of the system volume is performed by depicting the mass measured by the MSB as a function of gas density ($\rho_f$) either using an equation of state (EOS) or through direct measurements which are facilitated using a sinker of a known volume ($V_{\text{sinker}}$):

$$\rho_{\text{f}} = \frac{(\text{MP}_2 - \text{MP}_1)_{\text{vac}} - (\text{MP}_2 - \text{MP}_1)_{\text{P}}}{V_{\text{sinker}}}. \tag{3.1}$$

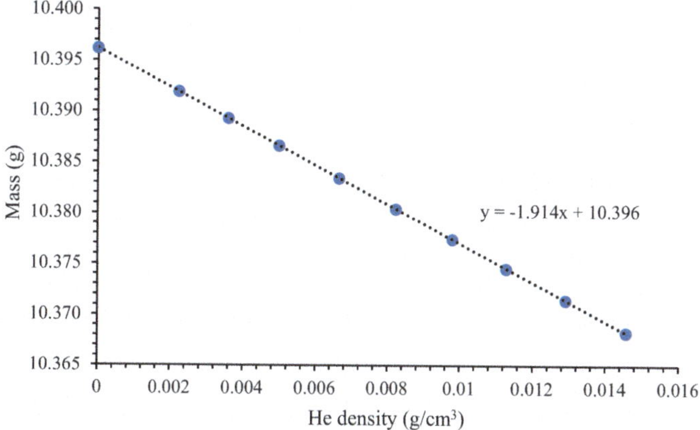

**Fig. 3.3**  Determination of system mass and volume (step 1)

A trend line can be fitted to the mass vs density graph, where the intercept represents the mass at vacuum, and the slope is the volume of the system as shown in Fig. 3.3.

In the second step, the sample is loaded into the sample crucible and the measuring cell is tightly closed and exposed to vacuum at elevated temperatures to remove any preadsorbed species. A stable measured mass with time indicates the activation and readiness of the sample for the consecutive step as shown in Fig. 3.4. This step is often referred to as a pretreatment step.

During the third step and similar to the first step, the mass and the skeletal volume of the system including the sample are determined using He. The difference between the mass obtained in the third step and the mass obtained in the first step is the mass

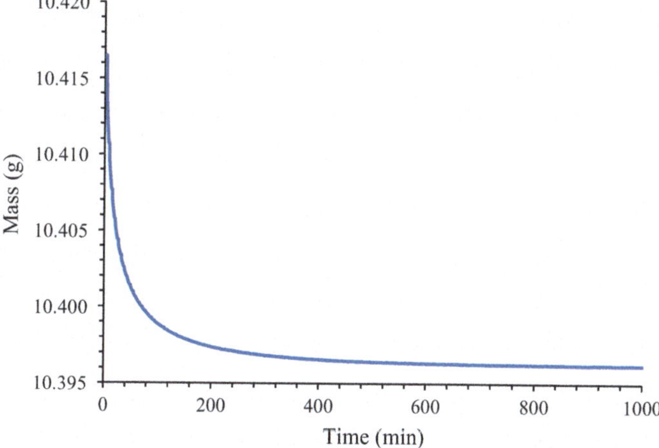

**Fig. 3.4**  Measured mass during the activation of a shale sample (step 2)

of the sample at vacuum. Similarly, the difference between the volume obtained in the third step and the first step is the volume of the sample. The relevance of knowing the sample mass at vacuum is its use as a reference value for calculating the sorption capacity of the rock, normally expressed in mg/g rock, mmol/g rock, $cm^3$/g rock or as a weight percentage (wt%). After this step, the system is vacuumed again to remove any remaining helium. The last step comprises the measurement of the amount of adsorbed gas as a function of pressure. During this step, the measurement cell is set to the desired temperature and $CO_2$ pressure is increased stepwise to the desired value. Mass, pressure and temperature are logged periodically. Stable values over time signal that equilibrium is achieved and that the measurement at the next desired pressure can be initiated.

### 3.1.4   Evaluation of the Data

There exist two forces acting on the components within the MSB, firstly the weight acting downward which is given by (Wu et al. 2019):

$$F_w = \left(M_{sys,vac} + M_{rock,vac} + M_a\right) \cdot g, \tag{3.2}$$

where $M_{sys,vac}$ is the mass of the system at vacuum as acquired during the blank test, $M_{rock,vac}$ is the mass of the rock at vacuum, $M_a$ is the mass of the adsorbed layer, also referred to as the absolute adsorbed mass, and $g$ is the gravitation acceleration constant. The other force is the buoyancy force, defined as the mass of gas displaced by the sum of volumes, and is given by:

$$F_b = \rho_f\left(V_{sys} + V_{rock} + V_a\right) \cdot g, \tag{3.3}$$

where $\rho_f$ is the density of the gas which is a function of pressure and temperature, $V_{sys}$, $V_{rock}$ and $V_a$ are the volume of the system, the volume of the rock and the volume of the adsorbed layer, respectively. Accordingly, the mass measured by the MSB ($M_{MSB}$) is the difference between the weight and the buoyancy force, both divided by the gravitational acceleration constant, which yields:

$$M_{MSB} = \left(M_{sys,\,vac} + M_{rock,vac} + M_a\right) - \rho_f\left(V_{sys} + V_{rock} + V_a\right). \tag{3.4}$$

Rearranging Eq. (3.4) yields:

$$M_a = M_{MSB} - M_{sys,\,vac} - M_{rock,vac} + \rho_f\left(V_{sys} + V_{rock} + V_a\right). \tag{3.5}$$

All the variables in Eq. (3.5) can be determined experimentally in accordance with the procedures described earlier, except for the volume of the adsorbed layer ($V_a$). At low pressures, the volume of the adsorbed layer can be assumed negligible and the absolute adsorbed mass can be easily determined. Challenges, however, arise at elevated pressures, where the volume of the adsorbed layer is expected to grow, thereby influencing the measured adsorbed mass. Due to the buoyancy forces acting on the volume of the adsorbed layer, the measurement results in an apparent mass, lower than the real mass adsorbed. Experimental determination of the adsorbed layer volume is not possible. For practicality, this problem is circumvented by introducing the concept of the Gibbs surface excess ($M_{ex}$) (Sircar 1999, 2001, 2019), which is defined using the following equation:

$$M_{ex} = M_a - \rho_f \cdot V_a. \tag{3.6}$$

Accordingly, the experimentally measured apparent adsorbed mass ($m_{ex}$) can be determined using:

$$M_{ex} = M_{MSB} - M_{sys,vac} - M_{rock,vac} + \rho_f \left( V_{sys,\,vac} + V_{rock,vac} \right). \tag{3.7}$$

Since the absolute mass adsorbed is also given by:

$$M_a = \rho_a \cdot V_a, \tag{3.8}$$

where $\rho_a$ is an average density of the adsorbed layer, which also cannot be experimentally measured. Substituting Eq. (3.8) into Eq. (3.6) yields:

$$M_{ex} = \rho_a \cdot V_a - \rho_f \cdot V_a. \tag{3.9}$$

According to the above equation, the excess mass is defined as the amount of gas residing within the adsorption volume in excess of that which would be present in the absence of a solid surface (Sircar 1999). The relation between the excess mass and the absolute mass can be obtained from Eq. (3.9) as:

$$M_{ex} = \rho_a \cdot V_a \left( 1 - \frac{\rho_f}{\rho_a} \right) = M_a \left( 1 - \frac{\rho_f}{\rho_a} \right). \tag{3.10}$$

By dividing the excess mass measured at each pressure point by the mass of the rock as measured at vacuum, the adsorption capacity $n_{ex}$ can be determined:

$$n_{ex} = \frac{1}{\overline{M}} \frac{M_{ex}}{M_{rock,vac}}, \tag{3.11}$$

where $\overline{M}$ is the molar mass of the gas (44.1 g/mol or mg/mmol for $CO_2$). By depicting the values against pressure, an adsorption isotherm can be constructed (e.g., Fig. 3.5).

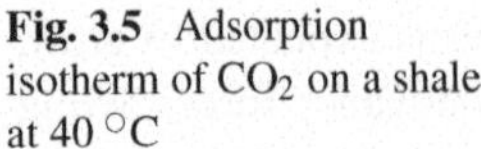

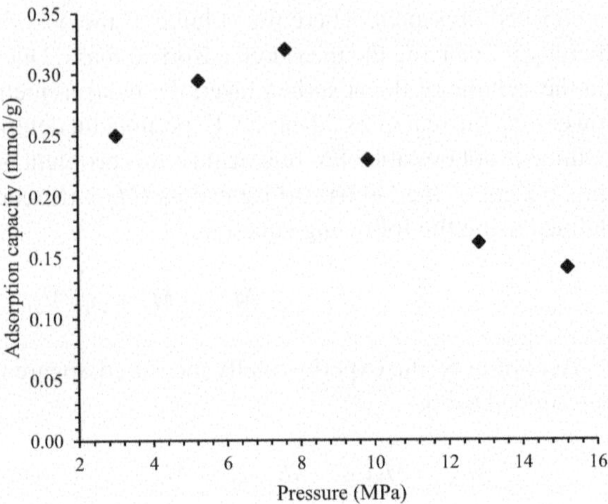

**Fig. 3.5** Adsorption isotherm of $CO_2$ on a shale at 40 $^\circ$C

## 3.2   The Volumetric Method

The volumetric method is the most commonly used gas sorption measurement technique due to its ease of use and maintenance as well as low cost. There are two configurations of the volumetric method: the constant pressure method (Mohammad et al. 2009) and the constant volume or the so-called manometric method (Gasparik et al. 2015).

### 3.2.1   The Experimental Setup

The setup of the manometric (Sieverts) method comprises a reference cell and a sample cell, both possessing precisely known volumes, separated by a valve operated by a computer- controlled electric actuator. Commonly, both cells are located within an air-heated or water-heated bath maintaining both at the same temperature. The setup also comprises a gas supply system and a vacuum pump. The schematic diagram of the manometric (constant volume) setup is shown in Fig. 3.6. In the constant pressure volumetric method, the reference cell is replaced by a varying volume device which maintains a constant pressure, i.e., a pump. The difference between the constant pressure and constant volume configurations is shown in Fig. 3.7. Experimental conditions are measured using high-precision pressure and temperature sensors, placed, preferably, in more than one location.

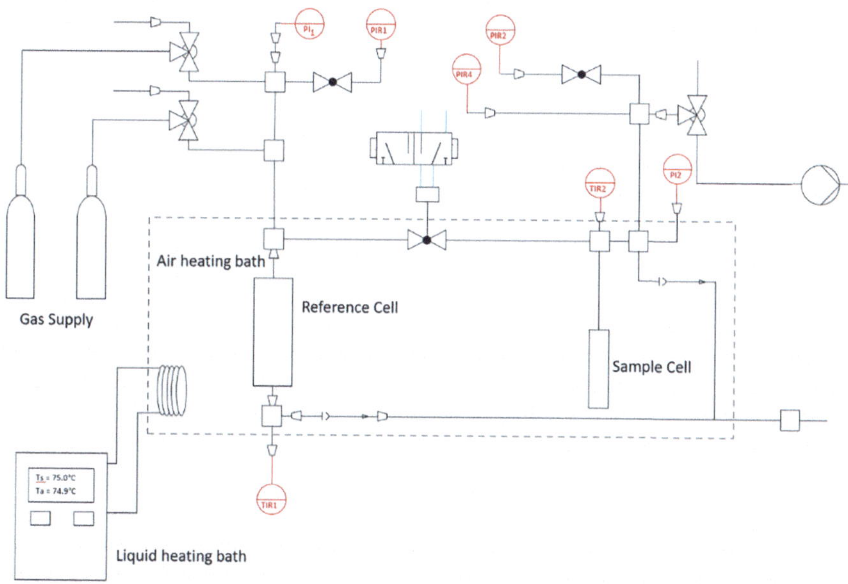


**Fig. 3.6** Schematic diagram of high-pressure high-temperature manometric sorption apparatus. Adapted from Eurotechnica GmbH. Used with permission

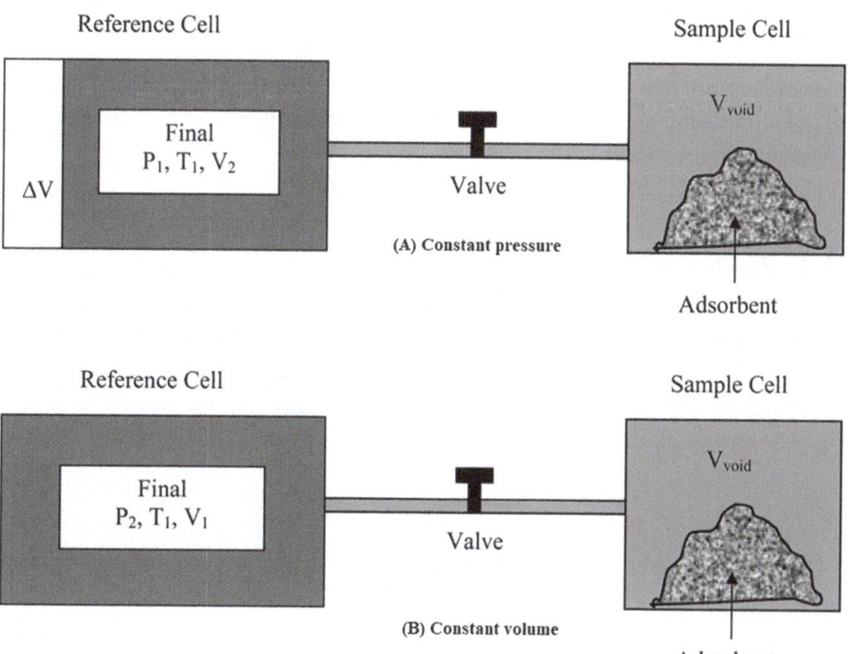


**Fig. 3.7** Difference between the constant pressure (**a**) and constant volume (**b**) configurations. Figure adapted from and modified after (Mohammad et al. 2009). Used with permission

### 3.2.2  Experimental Procedure

In analogy to the gravimetric method, volumetric gas sorption experiments are conducted in several steps. Initially, a certain amount of sample is weighed and loaded into the sample cell which is sealed and tightened. Vacuum is applied at elevated temperatures to remove any preadsorbed fluids. Afterward, multiple helium expansion tests are conducted at the operational conditions to determine the void volume ($V_{void}$) post sample introduction based on Boyle's law (Ross and Bustin 2007), much like in pycnometry. A consecutive step comprises applying vacuum again to remove residual helium. Afterward, the setup is heated to the desired experimental temperature to initiate the adsorption isotherm measurement. The gas of interest (e.g., $CO_2$) is introduced to the reference cell at a certain pressure. The values of pressure and temperature are continuously monitored and logged. After pressure and temperature equilibration within the reference cell, the valve separating the reference and sample cell is opened, after which gas expands into the sample cell. Pressure and temperature are also monitored and recorded during each gas expansion step. Any pressure drop occurring beyond the reason of gas expansion into a larger (precisely defined) volume is assumed to be due to gas adsorption (Broom 2011a). Stable values indicate equilibrium and that the next injection/expansion step can be initiated in a similar manner, i.e., the reference cell is loaded again with gas at a higher pressure, and after equilibration the gas is allowed to expand into the sample cell. In the case where a constant pressure configuration is applied, the valve separating the pump and the sample cell is opened and the piston of the pump moves to maintain a constant pressure (and therefore constant gas density) during injection and then the valve is closed after an appropriate mass of gas has been injected. Here, the volumes in the pump before and after injection are also recorded.

### 3.2.3  Evaluation of the Data

Both volumetric techniques rely on indirectly measuring the amount of adsorbed gas ($M_a$) by calculating the difference between the mass of injected gas from the reference cell (or pump) into the sample cell ($M_{inj,k}$) and the mass of free (unadsorbed) gas ($M_{free}$) (Gasparik et al. 2015):

$$M_a = M_{inj} - M_{free}. \tag{3.12}$$

The mass injected during several injection and expansion steps ($m$) is calculated by:

$$M_{inj,m} = \sum_{k=1}^{m} V_{RC}\left(\rho_{f,k} - \rho_{f,k+1}\right), \tag{3.13}$$

where $V_{RC}$ is the calibrated and known volume of the reference cell, $\rho_{f,k}$ is the density of the gas prior to gas injection into the sample cell and $\rho_{f,k+1}$ is the density of the gas after the expansion. The mass of free gas after expansion is determined by:

$$M_{\text{free}} = \rho_{f,k+1} V_{\text{void}}, \tag{3.14}$$

where $V_{\text{void}}$ is the volume in the sample cell accessible to helium (i.e., including sample pores). This volume also includes the volume within any pressure lines or fittings that connect the sample cell to the separating valve. The density of the gas can be determined using an appropriate equation of state (EOS) based on the measured pressure and temperature. The approach to determine $V_{\text{void}}$ is analogues to the determination of the sample volume in the gravimetric method. The evaluation is performed by depicting the amount of injected gas versus the helium density within the sample cell. The slope of the helium isotherm represents the void volume (Gasparik et al. 2014). After the determination of the void volume, the experiments can be conducted using the adsorptive gas at the desired temperature and an adsorption isotherm is then constructed by applying Eq. (3.12). The difference in this case would be that the drop in pressure after expansion is the result of both, volume filling and gas adsorption, as the adsorbed molecules do not contribute to the pressure measured within the sample cell.

Another method to construct the sorption isotherm of the sportive gas is by employing the real gas law, where the number of moles adsorbed at m injection step $m_{abs,m}$ is given by (Broom 2011a):

$$M_{a,m} = \sum_{k=1}^{m} \left( \frac{p_k V_{\text{RC}}}{Z_k RT} + \frac{p_{k-1} V_{\text{void}}}{Z_{k-1} RT} - \frac{p_{k+1}(V_{\text{void}} + V_{\text{RC}})}{Z_{k+1} RT} \right) \overline{M}, \tag{3.15}$$

where $p_k$ is the pressure prior to gas expansion, $p_{k+1}$ is the pressure after gas expansion and $p_{k-1}$ is the pressure measured in a previous step. $Z_k$ and $Z_{k+1}$ are gas compressibility factors before expansion of the gas and after, respectively, $Z_{k-1}$, is the compressibility factor corresponding to $p_{k-1}$. $\overline{M}$ is the molar mass of the gas. Accordingly, the second term on the right-hand side is the gas already available in the sample cell from a previous expansion step. For the first injection step ($m = 1$), the second term is zero as the pressure in the sample cell is at vacuum.

Much like the gravimetric method, the determination of the absolute adsorption using the volumetric method poses a challenge as the void volume changes, particularly at elevated pressures, due to a growing volume of the adsorbed phase. At low pressures, the void volume is given by:

$$V_{\text{void}} = V_{\text{f}}, \tag{3.16}$$

where $V_{\text{f}}$ is the volume of the free gas. Accordingly, at low pressure the mass adsorbed is calculated by:

$$M_a = M_{inj} - \rho_{f,void} V_f. \tag{3.17}$$

During the experiment and data evaluation, Eq. (3.17) is assumed to hold at higher pressures; however, at elevated pressures the void volume is given by:

$$V_{void} = V_f + V_a. \tag{3.18}$$

As such,

$$M_{free} = \rho_{f,\,void}(V_{void} - V_a). \tag{3.19}$$

And

$$M_a = M_{inj} - \rho_{f,void} V_{void} + \rho_{f,void} V_a. \tag{3.20}$$

As mentioned earlier $V_a$ cannot be determined experimentally. On the other hand, $M_{inj} - \rho_{f,void} V_{void}$ on the right-hand side of Eq. (3.20) corresponds to the apparent mass or excess mass, i.e.,

$$M_{inj} - \rho_{f,void} V_{void} = M_{ex}. \tag{3.21}$$

This equation is assumed to be valid at all pressures. Accordingly, Eq. (3.20) can be written as

$$M_{ex} = M_a - \rho_{f,void} V_a = M_a\left(1 - \frac{\rho_{f,void} V_a}{\rho_a V_a}\right) = M_a\left(1 - \frac{\rho_{f,void}}{\rho_a}\right). \tag{3.22}$$

This expression is the same as the expression of Eq. (3.10) for the gravimetric mass gain because $\rho_{f,void}$ is the same as $\rho_f$. Therefore, the volumetric and gravimetric methods should yield the same result regardless of the pressure level. By applying Eq. (3.11), the adsorption isotherm can then be constructed.


## 3.3  Measurement Uncertainties

The identification of sources of error is imperative for the accurate determination of adsorption capacity of rocks. In this section, a brief discussion on possible sources of uncertainty in the volumetric and gravimetric methods is presented.

### *3.3.1 The Volumetric Method*

As described earlier in the manometric method section, the experimental procedure consists of several injection steps from the reference cell into the sample cell. The mass injected at the $m$th injection/expansion step will be a cumulative sum of mass injected at all preceding steps (see Eq. 3.13). Accordingly, the uncertainty in the amount injected at the $m$th injection step is a cumulative sum of uncertainties of previously performed steps. In other words, this uncertainty accumulates. The uncertainty in the amount injected can stem from uncertainty in the volume of the reference cell (note that the material of the reference cell may mechanically expand when subjected to elevated pressures), the uncertainty in the density of the gas, or the compressibility factor which are on the one hand a result of uncertainty in the directly measured pressure and temperature and the uncertainty in the EOS applied to calculate these properties on the other. Therefore, it is imperative to use an accurate equation of state (e.g., Span and Wagner 1996) and pressure transducers with an uncertainty that renders negligible compared to the pressure drop expected to occur due to gas adsorption. The ability to detect the pressure drop will be influenced by the ratio of sample cell volume (and sample volume) to the reference cell volume. Depending on the pressure ranges intended for investigation, it is recommended to consider employing different pressure transducers with different measurement ranges to ensure sufficient accuracy. In terms of temperature, the uncertainty is usually negligible as the investigation of gas adsorption on rocks is conducted at temperatures corresponding to those that are common in the subsurface, i.e., far above ambient. However, the uncertainty in temperature measurement has a significant influence on the measured adsorption capacity particularly for $CO_2$ around its critical pressure, where any temperature change will result in a drastic change in $CO_2$ density (Di Giovanni et al. 2001). In any case, it is recommended to use high accuracy temperature sensors that are in direct contact with the sample, as temporal and spatial temperature variations may exist within the apparatus. One more source of uncertainty is the void volume (sample cell). This uncertainty is further magnified in experimental setups where the reference cell and the sample cell are maintained at different temperatures. In this case, the reference cell is maintained at a lower temperature to protect any temperature sensitive components, while the sample cell is kept at higher temperatures. As a result, a temperature gradient develops within the tubing connecting the reference cell and the sample cell. This gradient must be considered and quantified in the mass balance calculation (Gasparik et al. 2015). The presence of gas in cooler regions will induce a pressure drop that may be interpreted as adsorption and this uncertainty will be transferred to the calculated uptake (Broom 2011b). Furthermore, there is uncertainty whether the material of the sample cell mechanically and/or thermally expands under high pressure and/or high temperature conditions. Additionally, the determination of the sample mass at vacuum is challenging in the volumetric method, as degassing is usually performed after the sample is placed in the sample cell. The mass of the degassed sample may significantly differ due to the removal of preadsorbed gases or the presence of moisture and the change in mass will largely depend on the degassing duration. To reduce this uncertainty, the sample may be initially degassed in a separate vacuum oven before loading into the sample cell and the drop in mass should be monitored with time until no change can be observed. Care should also be taken during the weighing and the transfer of the sample into the sample cell; the contact of air (and therefore moisture) with the sample must be minimized to a high extent.

Leakage should be eliminated or reduced to the highest extent possible so that the mass balance calculations can be conducted without the need for a correction. A leakage test can be conducted prior to the experiment using helium. When a leak exists, the pressure drop will be linear with time, unlike adsorption which exhibits kinetic profiles of changing rates. A source of uncertainty may be the interaction of the gas with the material of the experimental setup, and whether gas adsorption occurs on the inner walls of the instrument's components.

### 3.3.2   The Gravimetric Method

The accuracy of the gravimetric method is generally higher in comparison to the volumetric method. The cumulative errors associated with the volumetric method are eliminated in the gravimetric method as the amount sorbed is measured directly at a given pressure and temperature instead of its determination through summing previously calculated quantities. Leakage does not affect the accuracy of the method as long as the pressure is kept constant by using a syringe pump, for example. Nevertheless, there are certain aspects that need to be taken into consideration for an accurate determination of the adsorption isotherm. Firstly, the resolution of the balance is of great significance as it should enable the detection of the slightest increase in mass during gas uptake. This will largely depend on the adsorption capacity of the rocks. A low resolution is particularly problematic in measuring gas adsorption on samples that are characterized with low capacity. The solution would be increasing the mass of the sample placed in the sample crucible. The sample mass, however, should not exceed the maximum allowable mass range provided by the manufacturer. Available commercial MSBs enable resolutions of 10 µg and sample mass range up to 25 g, rendering this source of uncertainty negligible. The accuracy of the gravimetric measurement is also reduced at elevated pressures where the influence of buoyancy on the microbalance reading is stronger. If the increase in sample weight is smaller than the increase in buoyancy force acting on the sample, the uncertainty increases. This could be particularly encountered in samples with relatively low adsorption capacities and in setups with small sample mass range (Gasparik et al. 2014).

Secondly, the accuracy of pressure and temperature measurement as well as EOS are of as much relevance in the gravimetric method as they are in the volumetric method. This is due to the fact that the density of the gas is required for buoyancy correction. Consequently, earlier discussions regarding the importance of an accurate EOS, temperature and pressure measurements still apply. For the accurate determination of the adsorption isotherm, the mass of the sample at the beginning of the experiment must also be well known. The mass can be measured directly from the balance at the end of the degassing phase. Ideally, at absolute vacuum, the mass should not be affected by the temperature as no gas molecules are present to induce buoyancy. However, absolute vacuum is impossible to achieve and the measured mass will depend on the temperature applied. Therefore, the mass is normally determined after degassing at a temperature that is similar to that intended for the actual adsorption isotherm. Nevertheless, it is also difficult to ensure that the sample temperature at vacuum is similar to the measured temperature as the temperature sensor is not in direct contact with the sample and a temperature gradient exists, resulting from the limited heat transfer at vacuum, where the thermal conductivity is extremely low (Broom 2011b). This results in uncertainty in the reference mass of the adsorbent. Moreover, the presence of the temperature sensor on the wall of the measuring cell causes deviations between the real temperature of the gas and the measured temperature, which may influence the buoyancy correction (Di Giovanni et al. 2001).

### 3.3.3  Other Considerations

A further issue to consider during an adsorption measurement is the presence of contaminants in the system. This can either be due to insufficient purity of the sorptive gas, the improper evacuation of the sample cell after helium expansion tests, contaminations in the apparatus, or the presence of moisture or light hydrocarbons in the sample which may elute during the adsorption isotherm measurement. This, on the one hand, influences the density of the gas, and on the other hand, the reference mass of the sample (measured at vacuum), which is necessary for determining the adsorption capacity. Equilibrium time must also be taken into account. Firstly, the Joule-Thompson effect causes the $CO_2$ to cool down during expansion. Therefore, it is imperative to ensure that thermal equilibrium is reached during each expansion or pressure increase step, which will ultimately influence the mechanical equilibrium (pressure) in the system. The equilibrium time will also be influenced by the degree to which the samples are crushed. Crushing enhances the connectivity of the pores and therefore smaller particles are expected to reach equilibrium faster (Chen et al. 2015; Hu et al. 2021).

## 3.4  Sample Volume and Void Volume Estimation Using Helium

Helium is conventionally used for determining skeletal sample volume in the gravimetric method and the void volume in the volumetric method. The use of helium is based on the assumption that it does not adsorb. Nonetheless, there have been studies where significant adsorption of helium is reported in nanoporous materials with narrow micropores (Brandani et al. 2016; Maggs et al. 1960; Malbrunot et al. 1997). Although elimination of helium adsorption can be principally achieved by conducting the expansion tests at elevated temperatures and low to moderate pressures (Gumma and Talu 2003; Malbrunot et al. 1997), other issues may arise, related to the postulation that gas accessibility and solid properties are independent of pressure and temperature. For instance, at low pressure, helium may not be able to enter the same set of micropores as it would at higher pressures where the penetrability of the gas could be enhanced (Li et al. 2019). Accordingly, the determination of volumes at low pressures results in either a larger skeletal sample volume for the gravimetric method or a smaller void volume in the volumetric method, both leading to the over-estimation of the excess adsorption according to Eqs. (3.7) and (3.21). Alternatively, determining the volume at a high pressure may result in helium adsorption, which translates to a larger void volume (Yu et al. 2020) and accordingly the underestimation of the excess adsorption (Ross and Bustin 2007). On the other hand, when the sample is exposed to elevated pressures, Ozdemir et al. (2003) argue that this may lead to sorbent shrinkage, i.e., the reduction in volume accessible to helium, and the creation of constrictions in pore throats, thereby prohibiting gas from entering pores that were initially accessible to helium.

The second complication arises from differences in the kinetic diameters of gas molecules. For instance, a helium molecule has a kinetic diameter of 0.26 nm (Kanezashi and Tsuru 2011) in comparison to $CO_2$, which possesses a kinetic diameter of 0.33 nm (Eguchi et al. 2012). This results in a size-exclusion effect, which poses a barrier to the diffusion of larger molecules and favors those with smaller kinetic diameter (Reid and Thomas 2001). As such, helium accesses larger volumes and the assumption that the sample volume in the gravimetric method or void volume in the volumetric method is the same during adsorption experiments no longer holds (Liu et al. 2024). Here we note that this comes in addition to a growing volume of adsorbed layer. Again, this overestimation of void volume will result in underestimating the adsorption capacity of rocks and such volume discrepancies require further calibrations through computer simulations (Liu et al. 2024; Yu et al. 2020). Another way to circumvent this problem would be the determination of the void volume or sample volume using $CO_2$ at pressure and temperature conditions at which adsorption does not occur, or is at least negligible. These conditions are nearly impossible to fulfill in the presence of micropores (Rouquerol et al. 2016). For example, in the gravimetric method, Nguyen et al. (2017) proposed applying the so-called subtraction method. Instead of relying on helium sample volume, the buoyancy correction is accounted for by conducting the adsorption experiment at similar conditions with the sorptive gas twice. The first run is conducted using the adsorbent, and the second is performed using a non-porous sample (silicon shot) possessing a negligible surface area and a similar volume as that of the adsorbent. In this case, the change in the non-porous sample mass with increasing pressure is ascribed only to buoyancy as gas adsorption on the silicon shot is assumed to be negligible. Subtracting the non-porous sample isotherm from the adsorbent isotherm at each pressure yields the excess amount adsorbed. Another possibility is to use an assumingly non-adsorbing gas of a similar kinetic diameter as that of the sorptive gas, e.g., argon (Ross and Bustin 2007), which has a kinetic diameter close to that of $CO_2$ (0.34 nm). However, this may also be challenging as it is unlikely that two different molecules possessing the same diameter would behave so differently in terms of adsorption. Lyu et al. (2021) and Rani et al. (2015) found that argon adsorbs substantially and report that helium is a better-suited gas for void volume determination.

## 3.5 The Challenge of Evaluating Adsorption Data

Under high pressures relevant to geological storage of $CO_2$, the density of $CO_2$ fluid will be high and the apparent mass gain (excess) in adsorption measurements could be much lower than the true mass gain (absolute). Moreover, the apparent mass gain could decrease with increasing $CO_2$ pressure after a peak value and could go down to zero and become negative, as shown in Figs. 3.8 and 3.9 for adsorption in shale and clay, respectively. In conventional reservoir rocks such as the Weyburn rock, $CO_2$ adsorption also showed such a pattern (Fig. 2.4). This behavior is the evidence that the apparent mass gains are not true adsorption values, which should not decrease with increasing pressure. Clearly, the measured adsorption cannot represent the adsorption capacity of these materials.

If the density of the adsorbed phase is known, the true adsorption $M_a$ can be determined from Eq. (3.22). However, the density cannot be measured. To evaluate the true adsorption from the excess adsorption, a common practice is by curve fitting. First, the true adsorption is assumed to follow a known pattern, such as the one

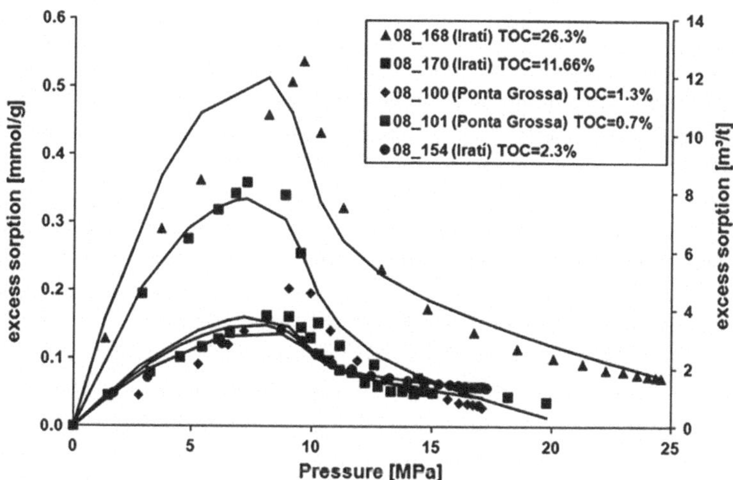


**Fig. 3.8** $CO_2$ excess adsorption on shale from Parana Basin in Brazil (Weniger et al. 2010)

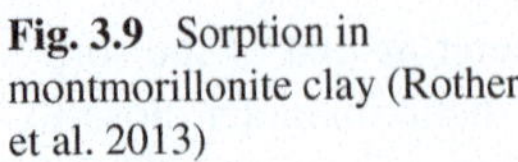

**Fig. 3.9** Sorption in montmorillonite clay (Rother et al. 2013)

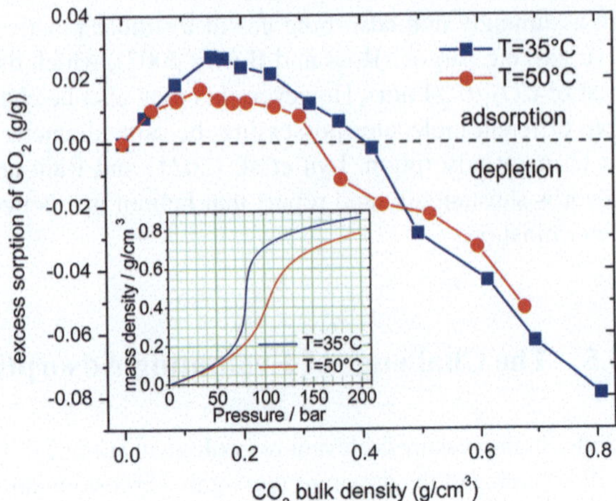


described by the Langmuir model that is introduced in Chap. 2:

$$C = \frac{C_{\mathrm{m}}bp}{1+bp},\qquad(2.3)$$

where $C$ is the amount of adsorbed gas per unit weight or volume of the adsorbent; $p$ is gas pressure; $C_{\mathrm{m}}$ and $b$ are parameters which are not dependent on pressure. By replacing $C$ in Eq. 2.3 with $M_{\mathrm{a}}/M_{\mathrm{r}}$, one obtains

$$\frac{M_{\mathrm{ex}}}{M_{\mathrm{rock}}} = \frac{C_{\mathrm{m}}bp}{1+bp}\left(1 - \frac{\rho_{\mathrm{f}}}{\rho_{\mathrm{a}}}\right).\qquad(3.23)$$

Then, the above equation is fitted to the measured isotherm of excess adsorption to determine the values of $\rho_{\mathrm{a}}$, $C_{\mathrm{m}}$ and $b$. However, the Langmuir model is based on several assumptions including that the adsorbed molecules form a single layer on the solid surface. Its validity for $CO_2$ adsorption in rocks has not been established. Moreover, there are three parameters ($\rho_{\mathrm{a}}$, $C_{\mathrm{m}}$ and $b$) to be determined and different sets of three parameter values could be used for the fitting. Under this circumstance, the fitting could be arbitrary because different sets of parameter values could give similar degree of fitting. Thus, the resulted estimate for the density could be far away from the true value. Furthermore, the Langmuir model cannot describe the data such as those shown in Fig. 3.10 for $CO_2$ adsorption in Weyburn rock.

To overcome the uncertainty associated with the density of adsorbed phase, a formula has been developed in this work for $CO_2$ storage which uses the measured apparent adsorption directly. This is described in the following chapter.

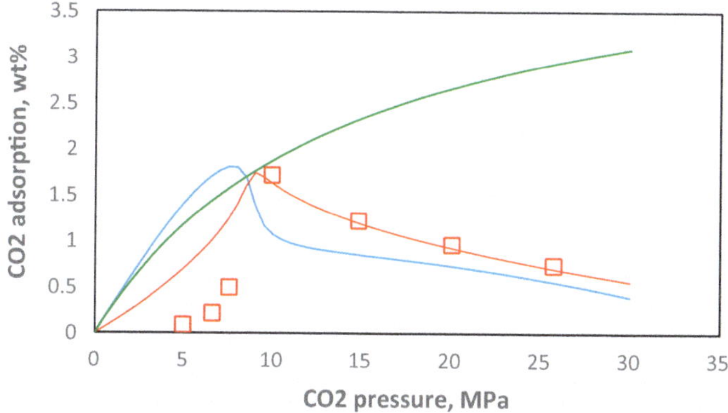



**Fig. 3.10** $CO_2$ adsorption behavior in Weyburn rock. The blue curve represents the fitting with the Langmuir model; the red curve represents the fitting with a modified Langmuir model (Wang et al. 2022). The green curve represents the actual adsorption derived from curve fitting.

# References


S. Brandani, E. Mangano, L. Sarkisov, Net, excess and absolute adsorption and adsorption of helium. Adsorption **22**, 261–276 (2016). https://doi.org/10.1007/s10450-016-9766-0

D.P. Broom, Gas Sorption Measurement Techniques, in *Hydrogen Storage Materials: The Characterisation of Their Storage Properties*. ed. by D.P. Broom (Springer, London, 2011), pp.117–139. https://doi.org/10.1007/978-0-85729-221-6_4

D.P. Broom, Experimental Considerations, in *Hydrogen Storage Materials: The Characterisation of Their Storage Properties*. ed. by D.P. Broom (Springer, London, 2011), pp.183–234. https://doi.org/10.1007/978-0-85729-221-6_6

Y. Chen, L. Wei, M. Mastalerz, A. Schimmelmann, The effect of analytical particle size on gas adsorption porosimetry of shale. Int. J. Coal Geol. **138**, 103–112 (2015). https://doi.org/10.1016/j.coal.2014.12.012

O. Di Giovanni, W. Dörfler, M. Mazzotti, M. Morbidelli, Adsorption of supercritical carbon dioxide on silica. Langmuir **17**, 4316–4321 (2001). https://doi.org/10.1021/la010061q

F. Dreisbach, H.W. Lösch, Magnetic suspension balance for simultaneous measurement of a sample and the density of the measuring fluid. J. Therm. Anal. Calorim. **62**, 515–521 (2000). https://doi.org/10.1023/A:1010179306714

R. Eguchi, S. Uchida, N. Mizuno, Highly selective sorption and separation of $CO_2$ from a gas mixture of $CO_2$ and $CH_4$ at room temperature by a zeolitic organic-inorganic ionic crystal and investigation of the interaction with $CO_2$. J. Phys. Chem. C **116**, 16105–16110 (2012). https://doi.org/10.1021/jp305890s

M. Gasparik, Y. Gensterblum, A. Ghanizadeh, P. Weniger, B.M. Krooss, High-pressure/high-temperature methane-sorption measurements on carbonaceous shales by the manometric method: experimental and data-evaluation considerations for improved accuracy. SPE J. **20**, 790–809 (2015). https://doi.org/10.2118/174543-PA

M. Gasparik, T.F.T. Rexer, A.C. Aplin, P. Billemont, G. De Weireld, Y. Gensterblum, M. Henry, B.M. Krooss, S. Liu, X. Ma, R. Sakurovs, Z. Song, G. Staib, K.M. Thomas, S. Wang, T. Zhang, First international inter-laboratory comparison of high-pressure $CH_4$, $CO_2$ and C2H6 sorption isotherms on carbonaceous shales. Int. J. Coal Geol. **132**, 131–146 (2014). https://doi.org/10.1016/j.coal.2014.07.010

S. Gumma, O. Talu, Gibbs dividing surface and helium adsorption. Adsorption **9**, 17–28 (2003). https://doi.org/10.1023/A:1023859112985

B. Hu, Y. Cheng, L. Wang, K. Zhang, X. He, M. Yi, Experimental study on influence of adsorption equilibrium time on methane adsorption isotherm and Langmuir parameter. Adv. Powder Technol. **32**, 4110–4119 (2021). https://doi.org/10.1016/j.apt.2021.09.015

M. Kanezashi, T. Tsuru, Chapter 6—Gas Permeation Properties of Helium, Hydrogen, and Polar Molecules Through Microporous Silica Membranes at High Temperatures: Correlation with Silica Network Structure, in *Membrane Science and Technology, Inorganic Polymeric and Composite Membranes.* ed. by S.T. Oyama, S.M. Stagg-Williams (Elsevier, Amsterdam, 2011), pp.117–136. https://doi.org/10.1016/B978-0-444-53728-7.00006-9

J.U. Keller, F. Dreisbach, H. Rave, R. Staudt, M. Tomalla, Measurement of gas mixture adsorption equilibria of natural gas compounds on microporous sorbents. Adsorption **5**, 199–214 (1999). https://doi.org/10.1023/A:1008998117996

R. Kleinrahm, X. Yang, M.O. McLinden, M. Richter, Analysis of the systematic force-transmission error of the magnetic-suspension coupling in single-sinker densimeters and commercial gravimetric sorption analyzers. Adsorption **25**, 717–735 (2019). https://doi.org/10.1007/s10450-019-00071-z

J. Li, K. Wu, Z. Chen, K. Wang, J. Luo, J. Xu, R. Li, R. Yu, X. Li, On the negative excess isotherms for methane adsorption at high pressure: modeling and experiment. SPE J. **24**, 2504–2525 (2019). https://doi.org/10.2118/197045-PA

B. Liu, S. Babaei, M. Kanduč, S. Tian, L. Bai, Y. Xu, M. Ostadhassan, Helium expansion revisited: effects of accessible volume on excess adsorption in kerogen matrices. Chem. Eng. J. **493**, 152690 (2024). https://doi.org/10.1016/j.cej.2024.152690

Y. Lyu, D. Dasani, T. Tsotsis, K. Jessen, Characterization of shale using helium and argon at high pressures. J. Pet. Sci. Eng. **206**, 108952 (2021). https://doi.org/10.1016/j.petrol.2021.108952

F.A.P. Maggs, P.H. Schwabe, J.H. Williams, Adsorption of helium on carbons: influence on measurement of density. Nature **186**, 956–958 (1960). https://doi.org/10.1038/186956b0

P. Malbrunot, D. Vidal, J. Vermesse, R. Chahine, T.K. Bose, Adsorbent helium density measurement and its effect on adsorption isotherms at high pressure. Langmuir **13**, 539–544 (1997). https://doi.org/10.1021/la950969e

D.L. Minnick, T. Turnaoglu, M.A. Rocha, M.B. Shiflett, Review Article: Gas and vapor sorption measurements using electronic beam balances. J. Vac. Sci. Technol. Vac. Surf. Films **36**, 050801 (2018). https://doi.org/10.1116/1.5044552

S. Mohammad, J. Fitzgerald, R.L. Robinson, K.A.M. Gasem, Experimental uncertainties in volumetric methods for measuring equilibrium adsorption. Energy Fuels **23**, 2810–2820 (2009). https://doi.org/10.1021/ef8011257

H.G.T. Nguyen, J.C. Horn, M. Thommes, R.D. van Zee, L. Espinal, Experimental aspects of buoyancy correction in measuring reliable high-pressure excess adsorption isotherms using the gravimetric method. Meas. Sci. Technol. **28**, 125802 (2017). https://doi.org/10.1088/1361-6501/aa8f83

E. Ozdemir, B.I. Morsi, K. Schroeder, Importance of volume effects to adsorption isotherms of carbon dioxide on coals. Langmuir **19**, 9764–9773 (2003). https://doi.org/10.1021/la0258648

S. Rani, B.K. Prusty, S.K. Pal, Comparison of void volume for volumetric adsorption studies on shales from India. J. Nat. Gas Sci. Eng. **26**, 725–729 (2015). https://doi.org/10.1016/j.jngse.2015.07.012

C.R. Reid, K.M. Thomas, Adsorption kinetics and size exclusion properties of probe molecules for the selective porosity in a carbon molecular sieve used for air separation. J. Phys. Chem. B **105**, 10619–10629 (2001). https://doi.org/10.1021/jp0108263

D.J.K. Ross, R. Bustin, Impact of mass balance calculations on adsorption capacities in microporous shale gas reservoirs. Fuel **86**, 2696–2706 (2007). https://doi.org/10.1016/j.fuel.2007.02.036

G. Rother, E.S. Ilton, D. Wallacher, T. Hauß, H.T. Schaef, O. Qafoku, K.M. Rosso, A.R. Felmy, E.G. Krukowski, A.G. Stack, N. Grimm, R.J. Bodnar, $CO_2$ sorption to subsingle hydration layer montmorillonite clay studied by excess sorption and neutron diffraction measurements. Environ. Sci. Technol. **47**, 205–211 (2013). https://doi.org/10.1021/es301382y

J. Rouquerol, F. Rouquerol, P. Llewellyn, R. Denoyel, Surface excess amounts in high-pressure gas adsorption: Issues and benefits. Colloids Surf. Physicochem. Eng. Asp. **496**, 3–12 (2016). https://doi.org/10.1016/j.colsurfa.2015.10.045

S. Sircar, The genius of Gibbsian surface excess (GSE) framework for fluid (gas or liquid)—solid adsorption: a powerful practical tool. Sep. Purif. Technol. **213**, 235–245 (2019). https://doi.org/10.1016/j.seppur.2018.11.070

S. Sircar, Measurement of gibbsian surface excess. AIChE J. **47**, 1169–1176 (2001). https://doi.org/10.1002/aic.690470522

S. Sircar, Gibbsian surface excess for gas adsorption revisited. Ind. Eng. Chem. Res. **38**, 3670–3682 (1999). https://doi.org/10.1021/ie9900871

R. Span, W. Wagner, A new equation of state for carbon dioxide covering the fluid region from the triple-point temperature to 1100 K at pressures up to 800 MPa. J. Phys. Chem. Ref. Data **25**, 1509–1596 (1996). https://doi.org/10.1063/1.555991

J. Wang, H. Samara, P. Jaeger, V. Ko, D. Rodgers, D. Ryan, Investigation for $CO_2$ adsorption and wettability of reservoir rocks. Energy Fuels **36**, 1626–1634 (2022). https://doi.org/10.1021/acs.energyfuels.1c03366

G.D. Weireld, M. Frère, R. Jadot, Automated determination of high-temperature and high-pressure gas adsorption isotherms using a magnetic suspension balance. Meas. Sci. Technol. **10**, 117–126 (1999). https://doi.org/10.1088/0957-0233/10/2/010

Weniger, P., Kalkreuth, W., Busch, A., Krooss, B.M., 2010. High-pressure methane and carbon dioxide sorption on coal and shale samples from the Paraná Basin, Brazil, in *Joint 61st ICCP/ 26th TSOP Meeting: Advances in Organic Petrology and Organic Geochemistry*, Int. J. Coal Geol. **84**, 190–205. https://doi.org/10.1016/j.coal.2010.08.003

T. Wu, H. Zhao, S. Tesson, A. Firoozabadi, Absolute adsorption of light hydrocarbons and carbon dioxide in shale rock and isolated kerogen. Fuel **235**, 855–867 (2019). https://doi.org/10.1016/j.fuel.2018.08.023

X. Yang, R. Kleinrahm, M.O. McLinden, M. Richter, The magnetic suspension balance: 40 years of advancing densimetry and sorption science. Int. J. Thermophys. **44**, 169 (2023). https://doi.org/10.1007/s10765-023-03269-0

X. Yu, J. Li, Z. Chen, K. Wu, L. Zhang, S. Yang, Effects of helium adsorption in carbon nanopores on apparent void volumes and excess methane adsorption isotherms. Fuel **270**, 117499 (2020). https://doi.org/10.1016/j.fuel.2020.117499

# Chapter 4
# Evaluating $CO_2$ Storage Potential Using Apparent Adsorption and Effective Density

As has been discussed, the measured (apparent) adsorption is lower than the true (absolute) adsorption. This makes evaluation of $CO_2$ storage potential difficult, particularly when the apparent adsorption decreases with increasing $CO_2$ pressure like the case of the Weyburn carbonate. To tackle this problem, a method has been developed which bypasses the unknown density of adsorbed phase that is required to determine the absolute level of adsorption (Wang et al. 2022). Using this method, $CO_2$ storage capacity of rocks can be calculated from directly measured apparent adsorption data, without the need for assumed mechanistic models and density values.

## 4.1 Contribution of Adsorption to Total $CO_2$ Retention Capacity in Pore Space

In geological reservoirs, injected $CO_2$ will displace reservoir water in the pore network, as described earlier. In the pore space taken up by $CO_2$, adsorption occurs to an extent depending on the strength of the molecular interactions between the solid surface and $CO_2$. In this situation, two $CO_2$ phases, an adsorbed phase and a free phase, would coexist. In order to estimate $CO_2$ storage capacity using the measured adsorption data, the mass fraction of the adsorbed phase is related to the total pore space occupancy. Accordingly, the pore space filled by $CO_2$ may be divided into two parts:

$$V_p = V_f + V_a, \tag{4.1}$$

where $V_p$ is the pore space occupied by $CO_2$; $V_f$ and $V_a$ are the volumes occupied by the free phase and the adsorbed phase, respectively. Whereas $V_p$ can be measured for given rock, $V_a$ and $V_f$ cannot be determined separately. The two values may be related to measurable quantities as described in the following. The mass of free-phase $CO_2$

in the pore space is

$$M_f = \rho_f V_f = \rho_f\left(V_p - V_a\right) = \rho_f V_p - \rho_f V_a = \rho_f V_p - \rho_f M_a/\rho_a, \tag{4.2}$$

where Eq. 4.1 is used for $V_f$. The total mass of $CO_2$ in the pore space is the sum of adsorbed $CO_2$ and free-phase $CO_2$

$$M_t = M_a + M_f = M_a + \rho_f V_p - \frac{\rho_f M_a}{\rho_a} = M_a\left(1 - \frac{\rho_f}{\rho_a}\right) + \rho_f V_p$$
$$= \rho_f V_p\left[1 + \frac{M_a}{\rho_f V_p}\left(1 - \frac{\rho_f}{\rho_a}\right)\right]. \tag{4.3}$$

The pore volume $V_p$ in the term $\frac{M_a}{\rho_f V_p}$ in Eq. (4.3) can be related to skeletal volume of the rock $V_r$ by the porosity of the rock

$$\varepsilon = \frac{V_p}{V_p + V_r}, \tag{4.4}$$

where $\varepsilon$ is the porosity. Accordingly,

$$V_p = \frac{\varepsilon}{1 - \varepsilon}V_r. \tag{4.5}$$

Using this relation and noting that $V_r = M_r/\rho_r$, where $\rho_r$ is the skeletal density of rock, we rewrite Eq. (4.3) into

$$M_t = \rho_f V_p\left[1 + \frac{\rho_r}{\rho_f}\frac{(1 - \varepsilon)}{\varepsilon}\frac{M_a}{M_r}\left(1 - \frac{\rho_f}{\rho_a}\right)\right]. \tag{4.6}$$

The component $\frac{M_a}{M_r}\left(1 - \frac{\rho_f}{\rho_a}\right)$ corresponds to $\frac{M_{ex}}{M_r}$, the apparent or excess mass gain in adsorption measurement according to Eq. (3.10) divided by the mass of rock. Substituting this into Eq. (4.6) yields

$$M_t = \rho_f V_p\left[1 + \frac{\rho_r}{\rho_f}\frac{(1 - \varepsilon)}{\varepsilon}\frac{M_{ex}}{M_r}\right] \tag{4.7}$$

which relates the total mass of $CO_2$ that can be stored in a given pore space $V_p$ to the fluid density $\rho_f$, rock skeletal density $\rho_r$, rock porosity $\varepsilon$ and apparent adsorption of $CO_2$ per unit mass of rock $\frac{M_{ex}}{M_r}$. This equation shows that adsorption increases $CO_2$ storage capacity by a factor $\left(1 + \frac{\rho_r}{\rho_f}\frac{(1-\varepsilon)}{\varepsilon}\frac{M_{ex}}{M_r}\right)$ and this factor increases with $\frac{M_{ex}}{M_r}$, the measured apparent adsorption. Without adsorption, the storage capacity in a given pore space would be $M_t = \rho_f V_p$. Thus, the $CO_2$ storage capacity can be estimated from measured apparent adsorption without knowing the absolute value of the adsorption.

When the free-phase density reaches the adsorbed phase density, $\frac{M_{ex}}{M_r} = 0$ according to Eq. (3.10) and the two phases of $CO_2$ will no longer be distinguishable. At this point, the amount of $CO_2$ stored in the pore space will become $\rho_f V_p$, the same value as for the case without adsorption. However, the density of the free phase would in this case take the much higher value of the adsorbed phase density $\rho_f = \rho_a$. This will be discussed soon.

## 4.2   A Quick View of the Effects of Adsorption on $CO_2$ Storage Capacity and Overpressure

With the experimental data for the Weyburn carbonate, the factor $\left(1 + \frac{\rho_r}{\rho_f} \frac{(1-\varepsilon)}{\varepsilon} \frac{M_{ex}}{M_r}\right)$ is calculated to be 1.29 for 20% rock porosity; that is, nearly 30% more $CO_2$ can be stored in pore space owing to the adsorption. This value is close to an estimate for $CO_2$ adsorption in a sandstone reservoir using a different method by Wan et al. (2018).

Without adsorption, a 29% increase of $CO_2$ mass in the pore space of the Weyburn rock would require the pressure to be raised from 10 to 18 MPa at 40 °C. Such a large overpressure would have an adverse impact on the integrity of the reservoir. On the contrary, a substantial reduction of the overpressure due to adsorption would be a gratifying outcome.

The factor 1.29 mentioned above is calculated at the peak value of apparent adsorption at 10 MPa. At higher pressures, the value of the factor $\left(1 + \frac{\rho_r}{\rho_f} \frac{(1-\varepsilon)}{\varepsilon} \frac{M_{ex}}{M_r}\right)$ will decrease but the total amount of stored $CO_2$ can still increase due to increased density of the free phase. This is illustrated in Fig. 4.1, where two cases, one with adsorption and the other without adsorption, are compared for the storage capacity in Weyburn rock. Because the density of free-phase $CO_2$ always increases with pressure, the total $CO_2$ storage capacity will increase with pressure. However, the contribution of adsorbed phase to the total $CO_2$ may decrease. This can be seen from the observed peak of the apparent adsorption on Weyburn rock and Eq. (4.7), which shows that the value of the factor $\left(1 + \frac{\rho_r}{\rho_f} \frac{(1-\varepsilon)}{\varepsilon} \frac{M_{ex}}{M_r}\right)$ decreases with decreasing value of apparent adsorption $\frac{M_{ex}}{M_r}$, suggesting a decreasing contribution of the adsorbed phase to the total $CO_2$ held in the pore space.

Without adsorption, $CO_2$ storage capacity in given pore space can only increase with increasing pressure. An increasing $CO_2$ pressure will increase the $CO_2$ leaking potential, as high overpressure tends to increase $CO_2$ penetration through caprock and may even jeopardize its mechanical integrity. Therefore, the pressure must be controlled for the security of $CO_2$ storage. As has been discussed, adsorption can decrease $CO_2$ pressure due to higher density of the adsorbed phase. This effect can also be evaluated with Eq. (4.7). Figure 4.2 shows the calculated reduction of

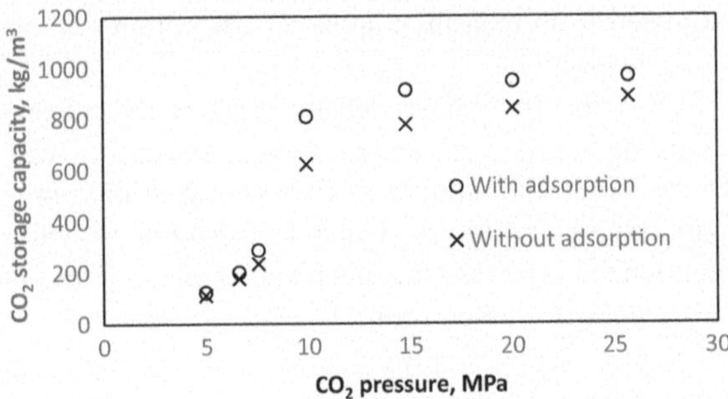



**Fig. 4.1** Calculated CO$_2$ storage capacity for Weyburn rock based on Eq. (4.7)

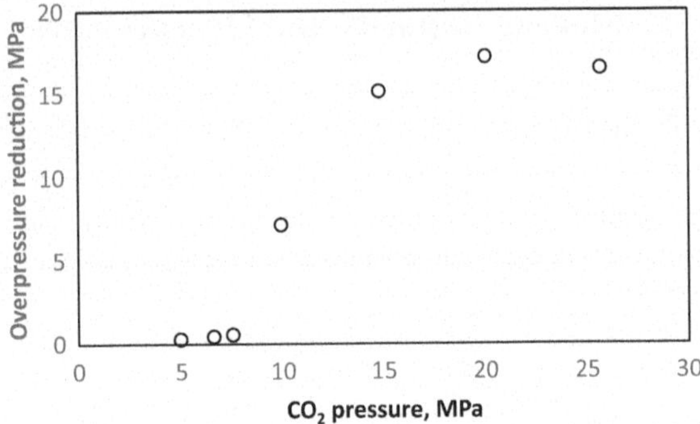



**Fig. 4.2** Calculated overpressure reduction due to CO$_2$ adsorption in Weyburn reservoir

overpressure owing to CO$_2$ adsorption for the Weyburn rock, where the overpressure is determined from the pressure dependence of CO$_2$ fluid density (see Fig. 2.2 in Chap. 2). It can be seen that the reduction of overpressure due to adsorption is substantial.

It should be emphasized again that using Eq. (4.7) only requires directly measured adsorption data $\frac{M_{ex}}{M_r}$, without the need of estimating the density of the adsorbed phase and the absolute adsorption. This is an important advantage of the approach.

For the Viking sandstone, the factor for increased CO$_2$ storage is even larger. The estimate is that over 40% more CO$_2$ can be stored than the case without CO$_2$ adsorption.

## 4.3  Excess Adsorption and Effective Density

The formula given by Eq. (4.7) conforms to the concept of excess adsorption. Whereas an adsorbed layer has a finite thickness, excess adsorption treats the adsorbed phase as a thickless layer. The three-dimensional space of the actual adsorbed layer is assumed to be occupied by unadsorbed gas phase. The difference between the adsorbed mass and the unadsorbed mass is the excess mass. In this way, the excess mass or excess adsorption can be viewed as absorbed or dissolved on the surface of the solid and not taking any space. Equation (4.7) treats the adsorbed $CO_2$ in an equivalent way. This can be seen by rewriting Eq. (4.7) into

$$M_t = \rho_f V_p + \rho_r V_p \frac{(1-\varepsilon)}{\varepsilon} \frac{M_{ex}}{M_r} = \rho_f V_p + \rho_r V_r \frac{M_{ex}}{M_r} = \rho_f V_p + M_{ex}, \qquad (4.7')$$

where Eqs. (4.4) and (4.5) have been used to convert $V_p$ to $V_r$ and to eliminate $\varepsilon$. Accordingly, the pore space is taken to be fully occupied by free-phase $CO_2$ with the mass $\rho_f V_p$; while the excess adsorption $M_{ex}$ contributes as an extra mass to $CO_2$ storage. Hence, Eq. (4.7) can describe $CO_2$ storage capacity in given pore space in terms of the excess adsorption and the density of the free phase.

Because adsorption increases $CO_2$ uptake in pore space, we can now apply the factor $\left(1 + \frac{\rho_r}{\rho_f} \frac{(1-\varepsilon)}{\varepsilon} \frac{M_{ex}}{M_r}\right)$ in Eq. (4.7) as an effective density coefficient and define an effective density as the product of this coefficient and the density of the free-phase $CO_2$, i.e.,

$$\rho_{eff} = \rho_f \left(1 + \frac{\rho_r}{\rho_f} \frac{(1-\varepsilon)}{\varepsilon} \frac{M_{ex}}{M_r}\right) \qquad (4.8)$$

where $\rho_{eff}$ is the effective density of $CO_2$ in adsorbed and free phases. The total amount of $CO_2$ that can be stored in given pore space is then expressed by

$$M_t = \rho_{eff} V_p. \qquad (4.9)$$

This formula can be used to estimate $CO_2$ storage capacity, be it region-wide or reservoir-wide, as long as the pore space available for $CO_2$ can be estimated. The effective density is a function of $CO_2$ pressure and adsorption capacity of rocks which can be evaluated experimentally. For example, for a given reservoir the $CO_2$ capacity can be estimated as

$$M_t = \rho_{eff} A h \varepsilon (1 - S_w), \qquad (4.10)$$

where $A$ and $h$ denote the area and the average thickness of the reservoir, respectively. $S_w$ represents water saturation of the rock which can be measured. With experimental data for $CO_2$ adsorption on the rock at the reservoir temperature and pressure and measured rock porosity and skeletal density, the effective density $\rho_{eff}$ can be determined using Eq. (4.8). The $CO_2$ storage capacity can then be estimated using Eq. (4.10).

We now show that the effective density is equivalent to the volume-weighted average density of the adsorbed phase and free-phase $CO_2$ in given pore volume, which is given by

$$\rho_{av} = \frac{\rho_a V_a + \rho_f V_f}{V_a + V_f} = \frac{\rho_a V_a + \rho_f V_f}{V_P}. \qquad (4.11)$$

This weighted average density cannot be determined without knowing the density and volume of the adsorbed phase. However, by replacing $V_f$ with $V_P - V_a$, Eq. (4.11) can be rearranged into

$$\rho_{av} = \frac{\rho_a V_a + \rho_f V_P - \rho_f V_a}{V_P} = \frac{\rho_f V_P + \rho_a V_a \left(1 - \frac{\rho_f}{\rho_a}\right)}{V_P}$$

$$= \rho_f \left[1 + \frac{\rho_r}{\rho_f} \frac{(1-\varepsilon)}{\varepsilon} \frac{M_{ex}}{M_r}\right] = \rho_{eff} \tag{4.12}$$

where Eq. (4.5) is used in the derivation to express $V_P$ in terms of other variables. Equation (4.12) shows that the effective density is an expression of the volume-weighted average density which explicitly shows the contribution of the adsorbed phase. Both expressions will be used for further evaluations for $CO_2$ storage.

The controlling factor for adsorption capacity of solids is the surface area. The larger the surface area, the higher the adsorption capacity. With given porosity, the surface area increases with decreasing pore size. This effect is not explicitly reflected in Eqs. (4.7) and (4.8). To see the effect of pore size, Eq. (4.7) can be rearranged into

$$M_t = \rho_f V_p \left[1 + \frac{2M_a}{\rho_f \bar{r} S_p} \left(1 - \frac{\rho_f}{\rho_a}\right)\right]$$

$$= \rho_f V_p \left[1 + \frac{2C_s(p) S_p}{\rho_f \bar{r} S_p} \left(1 - \frac{\rho_f}{\rho_a}\right)\right]$$

$$= \rho_f V_p \left[1 + \frac{2C_s(p)}{\rho_f \bar{r}} \left(1 - \frac{\rho_f}{\rho_a}\right)\right] \tag{4.13}$$

by replacing $V_p$ in Eq. (4.7) with $\bar{r} S_p / 2$, where $\bar{r}$ is the mean value of the pore radii and $S_p$ is the pore surface. This is based on the simplification $V_p = \pi r^2 l$ and $S_p = 2\pi r l$, where $l$ is the length of the pore. $C_s(p)$ is the surface concentration of adsorbed $CO_2$, which depends on $CO_2$ pressure and property of the rock (adsorption capacity). Consequently, the effective density of $CO_2$ can be determined by

$$\rho_{eff} = \rho_f \left[1 + \frac{2C_s(p)}{\rho_f \bar{r}} \left(1 - \frac{\rho_f}{\rho_a}\right)\right] \tag{4.14}$$

which increases with decreasing pore size. In tight rocks where average pore size is small, adsorption can be the dominating mechanism for $CO_2$ storage.

Without adsorption on pore surface, the amount of $CO_2$ stored in the pore space of unit reservoir volume would be

$$\frac{\rho_f V_p}{V_r + V_p} = \rho_f \varepsilon \tag{4.15}$$

which is dependent on porosity. With adsorption, the amount of $CO_2$ stored in the same reservoir volume is dependent on pore size as well. As can be seen from Eq. (4.14), smaller pores enable higher $CO_2$ storage capacity because of the adsorption of $CO_2$ on the pore surfaces. However, smaller pores result in higher capillary pressure which opposes $CO_2$ entering. Here the effects of $CO_2$ adsorption on changing the wettability of rock surfaces and reducing the capillary pressure become relevant, which will be discussed later.

## 4.4  Pressure Space and the Importance of Density and Overpressure Reduction

A concept of pressure space has been proposed recently in recognition of the fact that $CO_2$ storage is not only limited by the available pore volume but also limited by allowable overpressure (Bump and Hovorka 2024). The pressure space is defined as the mathematical product of accessible pore volume and allowable pressure increase in the reservoir. The allowable pressure increase depends on various factors, including the initial reservoir pressure, site-specific pressure limits, and the integrity of wells. Elevated pressure in the injection zone can increase safety and environmental risks, such as the potential for induced seismicity, compromised integrity of surrounding geological formations, and causing high-salinity brine to contaminate overlying drinking-water reservoirs which can be more harmful than the contamination by leaked $CO_2$ (Bump and Hovorka 2024).

Conventional evaluations of $CO_2$ storage focus on pore volume available for $CO_2$ injection and expected water saturation. The pressure space concept incorporates the factor of pressure increase associated with $CO_2$ injection. In particular, it recognizes that the elevated pressure, which goes beyond the area occupied by the $CO_2$ plume itself, can affect neighboring $CO_2$ injection projects. As such, the pressure footprint of a particular injection project is not bound by the $CO_2$ plume but can be much larger. When multiple injection projects coexist, the pressure space and hence the storage capacity will be lower. With the pressure limit, recent estimates suggest that pore volume utilization is below 1% (Bump and Hovorka 2024).

In view of the pressure limitation, it has been proposed to inject $CO_2$ at deeper locations. This strategy has several potential benefits:

1. As the reservoir pressure increases with depth, the rocks at greater depths can withstand higher injection pressure.
2. The higher pressure at greater depths can make $CO_2$ denser and decrease its density difference with reservoir brine, thereby decreasing the buoyancy force.
3. Higher density at greater depths will increase $CO_2$ storage capacity.
4. Greater depths will also prolong the time for $CO_2$ plume to reach the top seal. Longer travel time increases $CO_2$ dissolution and convective mixing, promoting solubility and mineral trapping.

However, great depths are not achievable in all reservoirs. Besides, the porosity of the reservoirs would decrease with increasing depth, which reduces available pore space. Moreover, the salinity of brine increases with depth, which decreases $CO_2$ solubility. Further, deep depth decreases the height of $CO_2$ column that can be withstood by caprocks (Iglauer 2018), thereby endangering the sealing capacity. In addition, deep wells increase drilling costs. Energy consumption and costs of facilities for compressing $CO_2$ to higher pressure will also increase. Therefore, the practicability of deep injection is limited and a project-specific optimum depth exists.

Another proposed approach is extraction of native brine from $CO_2$ storage reservoirs (see, e.g., Bourcier et al. 2011; Buscheck et al. 2016a, b). In this approach, $CO_2$ injection wells are first used to extract reservoir brine and subsequently used to inject $CO_2$. As a result, $CO_2$ injection pressure can be lower and risks related to overpressure are reduced. Moreover, $CO_2$ storage capacity can be increased by freeing the space occupied by brine. The produced brine may be reinjected into separate reservoirs (Buscheck et al. 2016a). The United States Department of Energy (DOE) supported field tests for extraction and reinjection of brine for managing reservoir pressure and $CO_2$ storage efficiency (McNemar et al. 2021; Klapperich 2022). This approach depends on the availability of separate reservoirs for disposal of the brine, which is not widely attainable.

Then, to increase $CO_2$ storage capacity and efficiency with limited pressure space, increasing $CO_2$ density is one way to address the problem. For discussion of the effect of density, the pressure space may be expressed as

$$pV = p\frac{M}{\rho} \qquad (4.16)$$

where $M$ and $\rho$ are the mass and density of $CO_2$, respectively. When $pV$ is fixed, $M$ is proportional to $\rho$. With conventional evaluation which do not consider adsorption, $M/\rho$ is fixed when pressure and pore volume are fixed. With adsorption, however, the storage capacity can increase because of the higher effective density shown by Eq. (4.8). Alternatively, lower pressure could be applied to store given quantity of $CO_2$. This will increase the security and safety of $CO_2$ storage, because the risk of damaging the seal and penetration of $CO_2$ through the seal are lowered. In consequence, for given pressure space, the storage efficiency can be increased.

Adsorption itself can decrease pressure buildup because the higher density of adsorbed $CO_2$ layer. Without adsorption, the injected $CO_2$ would take more space and compress the rock and pore water to a higher degree. This necessarily causes greater overpressure. Clearly, the storage capacity and overpressure are interrelated and need to be evaluated together. In this regard, adsorption of $CO_2$ on pore surface can play an indispensable role.

# References


W.L. Bourcier, T.J. Wolery, T. Wolfe, C. Haussmann, T.A. Buscheck, R.D. Aines, A preliminary cost and engineering estimate for desalinating produced formation water associated with carbon dioxide capture and storage. Int. J. Greenhouse Gas Control **5**, 1319–1328 (2011). https://doi.org/10.1016/j.ijggc.2011.06.001

A.P. Bump, S.D. Hovorka, Pressure space: the key subsurface commodity for CCS. Int. J. Greenhouse Gas Control **136**, 104174 (2024). https://doi.org/10.1016/j.ijggc.2024.104174

T.A. Buscheck, J.M. Bielicki, J.A. White, Y. Sun, Y. Hao, W.L. Bourcier, S.A. Carroll, R.D. Aines, Pre-injection brine production in $CO_2$ storage reservoirs: an approach to augment the development, operation, and performance of CCS while generating water. Int. J. Greenhouse Gas Control **2016**(54), 499–512 (2016a). https://doi.org/10.1016/j.ijggc.2016.04.018

T.A. Buscheck, J.A. White, S.A. Carroll, J.M. Bielicki, R.D. Aines, Managing geologic $CO_2$ storage with pre-injection brine production: a strategy evaluated with a model of $CO_2$ injection at Snøhvit. Energy Environ. Sci. **9**(4), 1504–1512 (2016). https://doi.org/10.1039/C5EE03648H

R. Klapperich, Developing and validating pressure management and plume control strategies in the Williston Basin through a brine extraction and storage test (BEST) (2022). https://netl.doe.gov/sites/default/files/netl-file. Accessed 23 Feb 2025

S. Iglauer, Optimum storage depths for structural $CO_2$ trapping. Int. J. Greenhouse Gas Control **77**, 82–87 (2018). https://doi.org/10.1016/j.ijggc.2018.07.009

A. McNemar, L. Myer, M. McKoy, D. Damiani, G. Bromhal, Developing and validating pressure management and plume control strategies, in *Proceedings of the 15th Greenhouse Gas Control Technologies Conference*, 15–18 March 2021 Abu Dhabi, UAE (2021). https://doi.org/10.2139/ssrn.3811770.

J. Wan, T.K. Tokunaga, P.D. Ashby, Y. Kim, M. Voltolini, B. Gilbert, D.J. DePaolo, Supercritical $CO_2$ uptake by nonswelling phyllosilicates. Proc. Natl. Acad. Sci. USA **115**(5), 873–878 (2018). https://doi.org/10.1073/pnas.1710853114

J. Wang, H. Samara, P. Jaeger, V. Ko, D. Rodgers, D. Ryan, Investigation for $CO_2$ adsorption and wettability of reservoir rocks. Energy Fuels **36**(3), 1626–1634 (2022). https://doi.org/10.1021/acs.energyfuels.1c03366

# Chapter 5
# Absolute Adsorption and Fraction of Adsorbed $CO_2$ in Pore Space

## 5.1 Evaluation of the Density of Adsorbed $CO_2$

In the above analyses for storage capacity in given pore space and the potential for overpressure reduction, the apparent adsorption data can be used without knowing the density of the adsorbed $CO_2$ layer. However, to analyze other effects of adsorption such as the buoyancy effect, the density of this layer needs to be known. In the following, estimation of the density of the adsorbed layer will be discussed.

When the measured adsorption isotherms show a peak or maximum and then intercept the pressure axis, such as the one shown in Fig. 3.9 in Chap. 3, the density of the adsorbed phase may be taken as that of the free phase at the pressure of the intercept. This would require measurements under very high pressure in order to reach the interception point.

One interesting observation is that under similar pressure levels some rocks do not show a peak in the apparent adsorption isotherms, such as the case of the Viking sandstone. This may be attributed to higher adsorption capacity of the rock. For lower-capacity rocks and similar materials like porous silica (Rother et al. 2012), the maximum capacity is reached under a lower pressure. Above this pressure, the apparent adsorption would decrease with increasing pressure because of the increasing density of the surrounding bulk $CO_2$. For higher-capacity rocks, such as the Viking sandstone, the adsorption increases with increasing pressure which over-weighs the effect of the increasing density of bulk $CO_2$. As a result, no peak will show up.

We consider two situations for the peak behavior. One is that the saturation has not been reached. In this case, adsorption still increases with increasing pressure, but the mass gain due to the adsorption is overweighed by the increasing buoyancy due to the increased density of bulk $CO_2$. The other situation is that the saturation is reached so further increasing pressure will only increase the buoyancy. In this situation, the apparent adsorption would decrease with increasing pressure linearly, directly following the density of the bulk $CO_2$.


J. Wang et al., *Carbon Dioxide Adsorption in Rock and Geological Storage of Carbon*, SpringerBriefs in Applied Sciences and Technology, https://doi.org/10.1007/978-3-031-90218-5_5

From the relation between the apparent (or excess) mass gain and the absolute mass gain

$$M_{\text{ex}} = M_{\text{a}}\left(1 - \frac{\rho_{\text{f}}}{\rho_{\text{a}}}\right),\tag{3.10}$$

we can take the two mass gains $M_{\text{ex}}$ and $M_{\text{a}}$ as functions of the density of $CO_2$ fluid $\rho_{\text{f}}$. The density of adsorbed $CO_2$ $\rho_{\text{a}}$ is taken as constant in accordance with Polanyi's adsorption theory.

Differentiation of Eq. (3.10) yields

$$\begin{aligned}
\frac{\mathrm{d}M_{\text{ex}}}{\mathrm{d}\rho_{\text{f}}} &= \left(1 - \frac{\rho_{\text{f}}}{\rho_{\text{a}}}\right)\frac{\mathrm{d}M_{\text{a}}}{\mathrm{d}\rho_{\text{f}}} + M_{\text{a}}\frac{\mathrm{d}\left(1 - \frac{\rho_{\text{f}}}{\rho_{\text{a}}}\right)}{\mathrm{d}\rho_{\text{f}}} \\
&= \left(1 - \frac{\rho_{\text{f}}}{\rho_{\text{a}}}\right)\frac{\mathrm{d}M_{\text{a}}}{\mathrm{d}\rho_{\text{f}}} - \frac{M_{\text{a}}\mathrm{d}\rho_{\text{f}}}{\rho_{\text{a}}\mathrm{d}\rho_{\text{f}}} = \left(1 - \frac{\rho_{\text{f}}}{\rho_{\text{a}}}\right)\frac{\mathrm{d}M_{\text{a}}}{\mathrm{d}\rho_{\text{f}}} - \frac{M_{\text{a}}}{\rho_{\text{a}}}.
\end{aligned}\tag{5.1}$$

With the relation $\frac{M_{\text{a}}}{\rho_{\text{a}}} = V_{\text{a}}$, where $V_{\text{a}}$ is the volume of the adsorbed layer, we can write

$$\frac{\mathrm{d}M_{\text{ex}}}{\mathrm{d}\rho_{\text{f}}} = \left(1 - \frac{\rho_{\text{f}}}{\rho_{\text{a}}}\right)\frac{\rho_{\text{a}}\mathrm{d}V_{\text{a}}}{\mathrm{d}\rho_{\text{f}}} - V_{\text{a}} = (\rho_{\text{a}} - \rho_{\text{f}})\frac{\mathrm{d}V_{\text{a}}}{\mathrm{d}\rho_{\text{f}}} - V_{\text{a}}.\tag{5.2}$$

It can then be seen that, if the excess adsorption is plotted against the density of free-phase $CO_2$, the peak value will correspond to

$$(\rho_{\text{a}} - \rho_{\text{f}})\frac{\mathrm{d}V_{\text{a}}}{\mathrm{d}\rho_{\text{f}}} - V_{\text{a}} = 0,\tag{5.3}$$

i.e.,

$$\frac{\mathrm{d}V_{\text{a}}}{\mathrm{d}\rho_{\text{f}}} = \frac{V_{\text{a}}}{\rho_{\text{a}} - \rho_{\text{f}}}.\tag{5.4}$$

It should be noted that the peak values are not maximum values of absolute adsorption. At pressures beyond the peak, the adsorption layers still grow, but at a lower rate:

$$\frac{\mathrm{d}V_{\text{a}}}{\mathrm{d}\rho_{\text{f}}} < \frac{V_{\text{a}}}{\rho_{\text{a}} - \rho_{\text{f}}}.\tag{5.5}$$

When the actual adsorption reaches its maximum value, the adsorbed layer stops growing and $\mathrm{d}V_{\text{a}}/\mathrm{d}\rho_{\text{f}}$ becomes zero. With Eq. (5.3), we get

$$\frac{\mathrm{d}M_{\text{ex}}}{\mathrm{d}\rho_{\text{f}}} = -V_{\text{a}}^{\text{max}}.\tag{5.6}$$

That is, the slope of the isotherm takes a constant value, which is the negative of the maximum volume of the adsorbed layer. To see this behavior from a measured adsorption isotherm, we obtain from Eq. (5.6)

$$dM_{ex} = -V_a^{max} d\rho_f \tag{5.7}$$

and then

$$M_{ex} = -V_a^{max} \rho_f + C \tag{5.8}$$

where $C$ is a constant, whose value can be determined by introducing the following relation

$$M_{ex}^* = M_a^{max}\left(1 - \frac{\rho_f^*}{\rho_a}\right) = \rho_a V_a^{max}\left(1 - \frac{\rho_f^*}{\rho_a}\right) = V_a^{max}\left(\rho_a - \rho_f^*\right) \tag{5.9}$$

where $M_{ex}^*$ and $\rho_f^*$ are the excess adsorption and the fluid density values which correspond to the point where the maximum volume ($V_a^{max}$) of the adsorption layer occurs. Substituting Eq. (5.9) into Eq. (5.8), we obtain

$$C = \rho_a V_a^{max}. \tag{5.10}$$

From Eqs. (5.8) and (5.10), we arrive at

$$M_{ex} = V_a^{max}(\rho_a - \rho_f). \tag{5.11}$$

This equation describes the isotherm after the adsorption reaches the maximum value. In this portion of the isotherm, the apparent (excess) adsorption $M_{ex}$ decreases with increasing fluid phase density $\rho_f$ linearly. When the isotherm intercepts the $\rho_f$ axis, i.e., the apparent adsorption $M_{ex}$ becomes zero, the $\rho_f$ value will be equal to the density of adsorbed phase $\rho_a$. This is the value we need to evaluate the actual adsorption. If the isotherm does not cross the $\rho_f$ axis in the measured pressure range, $\rho_a$ can be evaluated from the linear equation which best fits the data points in the linear portion of the isotherm. As can be seen from Eq. (5.11), the linear equation will take the form

$$M_{ex} = \rho_a V_a^{max} - V_a^{max} \rho_f, \tag{5.11'}$$

where $\rho_a V_a^{max}$ is a constant which equals the maximum value of actual adsorption; $V_a^{max}$ is the coefficient of the variable $\rho_f$. The constant divided by the coefficient will yield the density of the adsorbed phase $\rho_a$.

The advantage of this method is that it can describe the measured isotherms directly without the need of a specific mechanistic model such as the Langmuir model. Whereas $\rho_a$ evaluated this way is dependent on the accuracy of measured adsorption data and could just serve as an estimate, it helps in evaluating several important

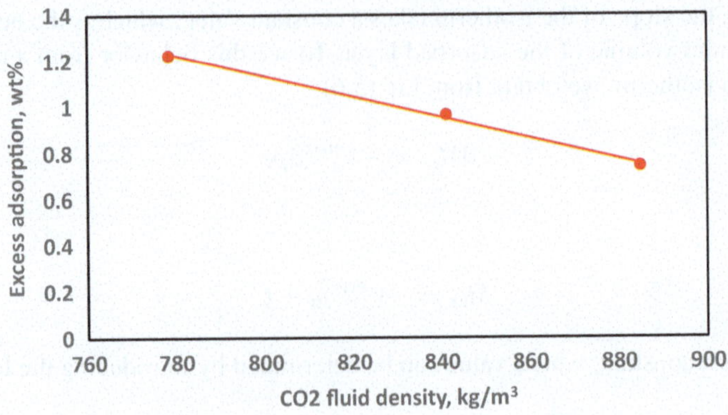


**Fig. 5.1** Evaluation of the density of adsorbed CO$_2$ phase from the dependence of excess adsorption on CO$_2$ density. The symbol represents adsorption data of Weyburn rock in the linear portion, and the line represents the fitting by Eq. (5.11′)

effects of CO$_2$ adsorption. With the above equation, a density value of 1050 kg/m$^3$ for adsorbed CO$_2$ phase in the Weyburn rock can be obtained, as illustrated in Fig. 5.1. This value agrees well with a value 1030 kg/m$^3$ obtained with a formula proposed by Dubinin (1960) (will be discussed shortly). However, Eq. (5.11′) can only be applied to evaluate the density of the adsorbed layer when the isotherm shows a peak. When the maximum adsorption capacity is not reached, a peak will not be observed in the measured pressure range. Nevertheless, we know that the true adsorption is higher than the apparent adsorption owing to the unaccounted volume of the adsorbed layer.

As mentioned earlier, the density of the adsorbed phase can be estimated from a relation proposed by Dubinin (1960) (Ozawa et al. 1976; Raveendran et al.2024). The relation is given by

$$d_a = 1/b \tag{5.12}$$

where $d_a$ is the molar density of the adsorbed phase and $b$ is the van der Waals constant appearing in the van der Waals equation of state for gases:

$$\left(p + \frac{a}{V_m^2}\right)(V_m - b) = RT, \tag{5.13}$$

where $p$ is the pressure of gas, $V_m$ is the molar volume of gas, $R$ is the universal gas constant and $T$ is the temperature of the gas. $a$ and $b$ are constants. $a$ reflects the interactions between gas molecules, and $b$ reflects the volume actually occupied by the molecules beside the empty space between the molecules. The value of $b$ for CO$_2$ is 0.0427 L/mol or 0.0427 m$^3$/kmol. The corresponding value of density calculated from the Dubinin relation is 1030 kg/m$^3$, which is quite close to the value 1050 kg/m$^3$ that is determined from the adsorption data of the Weyburn rock using Eq. (5.12), and close to a value 1080 kg/m$^3$ evaluated from adsorption data of coal (Siemens and Busch 2007).

The van der Waals equation is a well-known equation of state, which can be viewed as an extension of the ideal gas law

$$pV_m = RT \tag{5.14}$$

to describe the pressure–volume–temperature relationship of real gases. Unlike ideal gases where the molecules fill the container space $V_m$ but themselves are assumed not to occupy any space, real gas molecules occupy certain space which cannot be reduced. The free space that gas molecules can access is the total volume of the gas or the volume of the container minus the actual volume of gas molecules. With increasing pressure, the total volume of the gas decreases and the free space between gas molecules decreases, but when the molecules become tightly packed, the volume will not decrease further. This irreducible volume is considered to correspond to the constant $b$ in the van der Waals equation. The term $(V-b)$ is a correction to the term $V$ in the ideal gas law. By rearranging the van der Waals equation into

$$p = \frac{RT}{V_m - b} - a/V_m^2, \tag{5.13'}$$

we can see that when $V_m$ equals $b$, the pressure becomes infinitely large, suggesting that the molecules become incompressible like in a liquid or solid state. In the case of adsorption, the adsorbed phase is liquid-like according to Polanyi's theory. As such, the density of the adsorbed phase should be similar to that of the liquid. As the molar density equals the reciprocal of molar volume, the Dubinin relation Eq. (5.12) appears to be reasonable. One may envisage that the molecules of free-phase $CO_2$ are attracted to the pore surface, which results in compaction of the molecules to a liquid-like phase corresponding to the adsorbed state. The density of the adsorbed phase is equal to $1/b$ which is higher than the density of the free phase.

The values of the van der Waals constants are determined by applying to the equation the critical temperature condition $T = T_c$. As shown in Fig. 5.2, at the critical temperature the inflexion point corresponds to $\left(\frac{\partial p}{\partial V_m}\right)_{T_c} = 0$ and $\left(\frac{\partial^2 p}{\partial V_m^2}\right)_{T_c} = 0$. Under these conditions, the van der Waals equation yields

$$\left(\frac{\partial p}{\partial V_m}\right)_{T_c} = \frac{-RT}{(V_m - b)^2} + \frac{2a}{V_m^3} = 0 \tag{5.15}$$

$$\left(\frac{\partial^2 p}{\partial V_m^2}\right)_{T_c} = \frac{2RT}{(V_m - b)^3} - \frac{6a}{V_m^4} = 0 \tag{5.16}$$

Together with the van der Waals equation itself, the values of $a$ and $b$ can be determined as

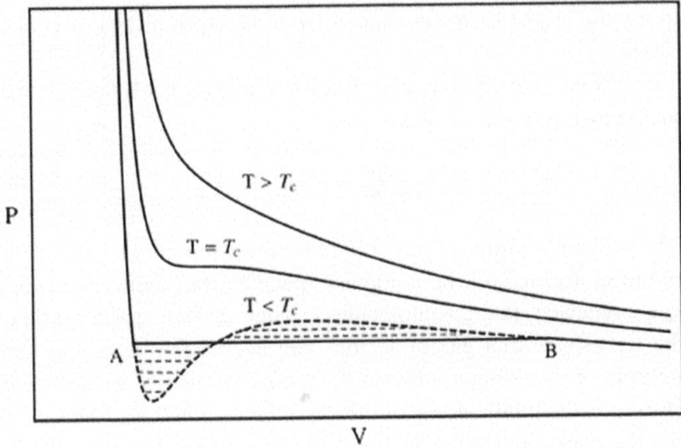



**Fig. 5.2** Pressure dependence of fluid volume in terms of the van der Waals equation (image from Moro 2014)

$$b = \frac{1}{3}V_{mc} \tag{5.17}$$

$$a = 3p_c V_{mc}^2, \tag{5.18}$$

where $V_{mc}$ and $p_c$ are the molar volume and pressure at critical temperature, respectively. These values can be determined experimentally for specific fluids, and the van der Waals constants are therefore fixed values. It is remarkable that the $b$ value, which is determined from $pVT$ data of supercritical fluids, could fit the adsorption data so well. It also suggests that the maximum CO$_2$ storage capacity would be

$$M_t = \rho_{eff} V_p = \frac{V_p}{b} = 1030 V_p \tag{5.19}$$

if the whole pore space is filled by adsorbed CO$_2$ and the adsorbed phase is incompressible. Under storage conditions, CO$_2$ is a supercritical fluid. Its density behavior can be described by an equation of state, which provides a relationship between the density, pressure and temperature. The van der Waals equation of state encompasses both gas and liquid states. With the aid of this equation, one can understand better the relationship between CO$_2$ density and pressure, two predominantly important factors for CO$_2$ storage, and the impact of adsorption which enables a higher-density CO$_2$ phase. In accordance with the Polanyi theorem, an adsorbed phase is liquid-like and the adsorption process is analogues to gas condensation. Whereas, unlike a liquid, the adsorbed phase is bound to the solid surface and not flowable, the density behavior of the fluid phase upon pressurization can be seen from the van der Waals equation. The equation describes the $pVT$ relations encompassing the critical point and condensation region, with well evaluated and documented constants.

The two totally different methods that have been used to estimate the density of the adsorbed $CO_2$ phase, one based on experimental data and one based on the value of the van der Waals constant $b$, yield close results which suggest the validity of the estimate. For more conservative assessments concerning the effects of the density of adsorbed $CO_2$ phase, the lower value 1030 kg/m$^3$ will be used to show to which extent $CO_2$ adsorption is able to decrease $CO_2$ buoyancy (and hence decrease the risk of $CO_2$ leakage) and contribute to $CO_2$ immobilization.

## 5.2  Absolute Adsorption

With the value 1030 kg/m$^3$ based on the van der Waals constant $b$, the absolute adsorption for the Weyburn rock is calculated by rearranging Eq. (3.10) in Chap. 3 into

$$M_a = \frac{M_{ex}}{\left(1 - \frac{\rho_f}{\rho_a}\right)}. \tag{3.10'}$$

Figure 5.3 shows a plot of the calculated absolute adsorption on the Weyburn rock as a function of $CO_2$ pressure. It can be seen that the absolute adsorption is much higher than the apparent adsorption when the pressure is higher than the critical pressure 73.8 bar. Moreover, above the pressure corresponding to the peak of the apparent adsorption, the absolute adsorption is still increasing, although at a slow rate. The pressure dependence of the actual adsorption has a similar trend to the pressure dependence of free-phase $CO_2$ density. Some implications can be drawn from the above behavior: First, the buoyancy force of high-density supercritical $CO_2$ phase results in the low apparent adsorption value. The adsorbed phase is separated from the fluid phase and is attached to rock sample. Second, the adsorption is significant enough to cause the large discrepancy between the apparent adsorption and the absolute adsorption. Third, $CO_2$ bulk density rather than pressure is the controlling factor for the adsorption, which consistently describes the trend of the absolute adsorption, as shown by a plot of the absolute adsorption against $CO_2$ fluid density in Fig. 5.4. Higher bulk density will bring about higher adsorption level. As the density of the adsorbed phase is taken as constant, it can be envisaged that a denser $CO_2$ fluid will be coexisting with a thicker adsorbed $CO_2$ layer.

## 5.3  Fractions of Adsorbed Phase and Free Phase in Pore Space

With the density value of the adsorbed phase, the volume of the adsorbed phase can be determined from

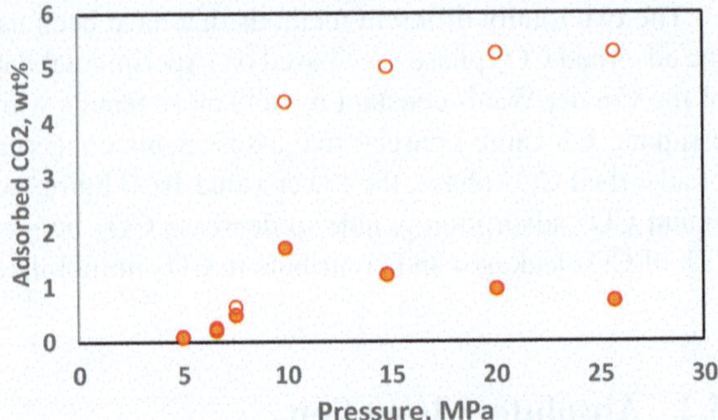

**Fig. 5.3** Evaluated absolute CO$_2$ adsorption in Weyburn rock. Solid symbol represents apparent (excess adsorption) and open symbol represents absolute adsorption

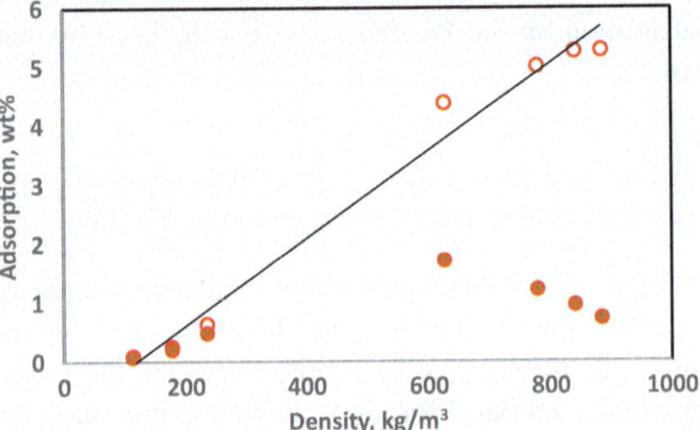

**Fig. 5.4** Dependence of CO$_2$ adsorption on CO$_2$ bulk density. Solid symbol represents apparent (excess adsorption), and open symbol represents absolute adsorption

$$V_a = \frac{M_a}{\rho_a} = \frac{M_{ex}}{\rho_a - \rho_f}. \tag{5.20}$$

The volume of the free-phase CO$_2$ can then be determined as

$$V_f = V_p - V_a = V_p - \frac{M_{ex}}{\rho_a - \rho_f}. \tag{5.21}$$

The volume fractions of the two phases in the pore space are therefore

$$\frac{V_a}{V_p} = \frac{M_{ex}}{(\rho_a - \rho_f)V_p} \tag{5.22}$$

and

$$\frac{V_f}{V_p} = 1 - \frac{M_{ex}}{(\rho_a - \rho_f)V_p}, \tag{5.23}$$

respectively. However, with the relation between the effective density and the volume-weighted average density discussed in Chap. 4, simpler formulas can be derived.

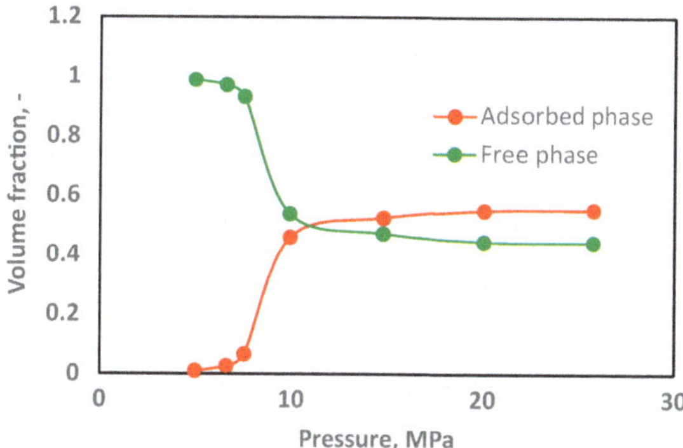

**Fig. 5.5** Volume fraction of adsorbed CO$_2$ and free-phase CO$_2$ in pore space of Weyburn rock



According to Eqs. (4.11) and (4.12), we can get

$$\rho_{av}V_P = \rho_a V_a + \rho_f V_f = \rho_{eff}V_P \tag{5.24}$$

and then

$$\frac{V_a}{V_p} = \frac{\rho_{eff} - \rho_f}{\rho_a - \rho_f} \tag{5.25}$$

$$\frac{V_f}{V_p} = \frac{\rho_a - \rho_{eff}}{\rho_a - \rho_f}. \tag{5.26}$$

Thus, the fraction of the adsorbed phase increases whereas the fraction of the free phase decreases with increasing effective density of CO$_2$ in the rock. Using these formulas, the volume fractions of the two phases in the pore space of Weyburn rock are calculated from the adsorption data and plotted against CO$_2$ pressure (Fig. 5.5). It can be seen that the fraction of adsorbed phase increases with CO$_2$ fluid pressure and exceeds the fraction of the free phase in the pressure range above 10 MPa. These results are important for CO$_2$ storage evaluations, because the two fractions correspond to the CO$_2$ that is immobilized by adsorption and the CO$_2$ that remains mobile, respectively. These results will be used to evaluate the buoyancy of CO$_2$ and its effect on CO$_2$ migration, which have a series of consequences for CO$_2$ storage.

## 5.4 Impact on Buoyancy of CO$_2$

In the previous chapter, we mentioned the effect of CO$_2$ adsorption on overpressure reduction in confined pore space. CO$_2$ adsorption can also reduce the overpressure caused by the buoyancy effect of the CO$_2$ plume, thereby reducing the potential for CO$_2$ leakage. This is due to reduced volume of mobile CO$_2$.

The caprocks of the reservoirs are relied upon to contain the $CO_2$ that tends to move upwards under the buoyancy force. The caprocks also contain pores, only fewer and smaller in size compared to the reservoir rock. If the caprocks are water-wet, the pores are filled with water and the associated capillary forces could prevent the $CO_2$ fluid from entering the pores. For $CO_2$ fluid to enter the pores and move up, the pressure of the $CO_2$ phase must be high enough to overcome the capillary forces. Once this happens, $CO_2$ fluid can penetrate the caprock and leak into the zones above the caprock.

When the injected $CO_2$ fluid reaches the caprock, the pressure at the top of the fluid phase will be higher than the pressure of the surrounding water due to the buoyancy force. This causes additional pressure on the caprock above the $CO_2$ fluid. The total force exerted by the $CO_2$ fluid on the caprock increases with the volume of the fluid. As $CO_2$ adsorption on the rock pore surface reduces the volume of free-phase $CO_2$, the upthrust on the caprocks decreases. Without adsorption, the volume of the $CO_2$ fluid will be larger and the upthrust on the caprocks will be greater, increasing the risk of $CO_2$ leakage. Moreover, the pressure on the caprock depends on the height of the fluid column. When the height increases, the pressure increases even if the total volume of $CO_2$ remains the same. The buoyancy force exerted by a fluid on an object submerged in the fluid is given by:

$$F_b = \rho V g, \qquad (5.27)$$

where $F_b$ is the buoyancy force; $\rho$ is the density of the fluid; $V$ is the volume of the fluid that is displaced by the submerged object; $g$ is the gravitational acceleration. In the case of a $CO_2$ column under the caprock, three forces are involved: the upward force from reservoir water, the gravitational force of the $CO_2$ column and the force from the caprock which counters the upthrust of $CO_2$. When the three forces are balanced, there is

$$F_c + \rho_f V_f g - \rho_w V_f g = 0, \qquad (5.28)$$

where $F_c$ is the downward force from the caprock; $\rho_f$ is the density of $CO_2$; $V_f$ is the volume of the $CO_2$ fluid; $\rho_w$ is the density of reservoir water. By replacing $V_f$ in Eq. (5.28) with a product $Ah$, where $A$ represents the average cross-sectional area of the column and $h$ represents the column height, we can rewrite Eq. (5.28) into

$$F_c = (\rho_w - \rho_f)Ahg. \qquad (5.29)$$

The pressure from the caprock will be

$$p = \frac{F_c}{A} = (\rho_w - \rho_f)gh. \qquad (5.30)$$

This pressure is equal to the pressure that the $CO_2$ column exerts on the caprock, i.e., the buoyancy pressure. The water surrounding the $CO_2$ column does not create a buoyancy pressure. As can be seen from the pressure gradients of an oil column in Fig. 5.6, which is analogous to a CO$_2$ column; at the bottom of the column the pressure is the same on CO$_2$ and water; while at the interface with the caprock the water pressure decreases by an amount $\rho_w gh$ whereas the pressure in the column decreases by an amount $\rho_f gh$. The difference $(\rho_w - \rho_f)gh$ is the pressure on the caprock which is an addition to the hydrostatic pressure at that depth due to the buoyancy effect. The fact that the additional pressure does not exist above the surrounding water is also implied by Eq. (5.30), which shows that when the density of the fluid column equals the density of water the buoyancy pressure becomes zero. Alternatively, a water column with the same height of the CO$_2$ column will not create a buoyancy pressure acting on the caprock.

In the above formulas, the pressure dependence of CO$_2$ density is neglected. Hence, the upthrust on the caprock by the upward buoyancy force is seen to be proportional to the volume of the fluid phase and the buoyancy pressure is seen to be proportional to the height of the column. When the column height is large and the density change due to the pressure change is significant, the buoyancy pressure can be given as

$$p = \left[\rho_w - \rho_f(\overline{p})\right]gh, \tag{5.31}$$

where $\rho_f(\overline{p})$ is a density value at an average pressure. The buoyancy pressure increases from the bottom of the CO$_2$ column to the top. As the buoyancy pressure is higher than the water pressure at the same depth, it can enable CO$_2$ to displace water from surrounding pores and adsorb on pore walls. As a result, the volume of mobile CO$_2$ will decrease as the adsorbed CO$_2$ is immobilized by its adherence to the solid surface. Consequently, the upward force on the caprock will decrease. Moreover, the height of the column could also decrease following CO$_2$ adsorption because of the decreased

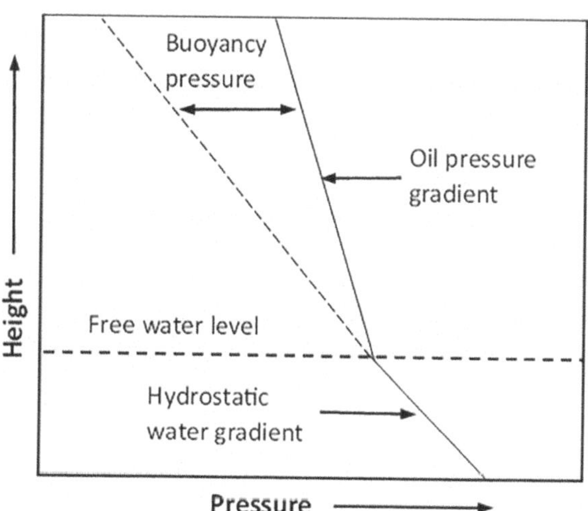


**Fig. 5.6** Illustration of density gradient and buoyancy pressure

volume of the free-phase $CO_2$. For understanding this effect we consider a column formed by a given mass of $CO_2$. The adsorbed portion in the column will increase with time because the top part of the column experiences higher buoyancy pressure, which causes expansion of mobile $CO_2$ into surrounding pores by overcoming the capillary pressure. This expansion increases $CO_2$ occupancy in pore space and as a result increases $CO_2$ adsorption on the pore surface. As the adsorbed phase has higher density, the volume of the $CO_2$ mass decreases and the pressure at the bottom of the $CO_2$ column decreases too. This will cause water to push the bottom of the column up, thereby decreasing the column height. In consequence, not only the total upthrust force but also the buoyancy pressure of the $CO_2$ column can be reduced owing to $CO_2$ adsorption. In the case of the Weyburn rock, the pressure on the caprock can be reduced by more than a half when the depth exceeds 1000 m, because more than a half of the total $CO_2$ will be immobilized by adsorption (see Fig. 5.5) and no longer contribute to the buoyancy force. This can largely decrease the risk of breaking the top sealing for $CO_2$.

It is worth noting that the adsorbed $CO_2$ phase receives little lifting force from water because its density ($> 1000$ kg/m$^3$) is close to that of the reservoir water. The lifting force, which is equals to $(\rho_w - \rho_a)g$, is thus negligible. This density effect, in addition to the adhesion to the rock surface, increases the tendency of adsorbed $CO_2$ phase to stay immobilized in the reservoir.

The buoyancy pressure at the top of the reservoir increases the driving force for $CO_2$ to spread laterally, which increases the area that needs to monitor as the potential of $CO_2$ leakage increases.

On the other hand, the $CO_2$ is more likely to interact with and adsorb to pore surfaces because the buoyancy pressure helps overcome the resistance to enter narrow pore spaces. The subsequent adsorption will decrease the mobility and increase $CO_2$ retention. The effective density of $CO_2$ can reflect the effect of adsorption on reducing the pressure on the caprock and the leakage risk zone. Toward the caprock the effective density would increase with adsorption, which translates to decreasing the fraction of mobile $CO_2$. As a result, the $CO_2$ plume boundary would decrease and the pressure decrease too. Thus, adsorption could reduce the footprint of $CO_2$ storage and reduce $CO_2$ leaking potential through reducing the buoyancy effect and the buoyant fraction of $CO_2$.

If $CO_2$ plume encounters structural or stratigraphic traps, adsorption will increase the trapping capacity because of the higher effective density. When no structural or stratigraphic traps exist, the buoyant force will maintain the pressure to push the $CO_2$ layer to spread laterally. Owing to adsorption, the free-phase $CO_2$ will decrease and the mobile layer will become thinner and thinner. As a result, the pressure of the mobile layer will decrease continuously. At last, either the mobile layer vanishes or the layer adheres to the surface of caprock. Thus, the reduction of the buoyancy force due to $CO_2$ adsorption can reduce the risk of $CO_2$ leakage.

# References


M.M. Dubinin, The potential theory of adsorption of gases and vapors for adsorbents with energetically nonuniform surfaces. Chem. Rev. **60**(2), 235–241 (1960). https://doi.org/10.1021/cr60204a006

A. Moro, Shock dynamics of phase diagrams. Ann. Phys. **343**, 49–60 (2014). https://doi.org/10.1016/j.aop.2014.01.011

S. Ozawa, S. Kusumi, Y. Ogino, Physical adsorption of gases at high pressure. IV. An improvement of the Dubinin-Astakhov adsorption equation. J. Colloid Interf. Sci. **56**(1), 83–91 (1976). https://doi.org/10.1016/0021-9797(76)90149-1

G. Raveendran, K. War, D.N. Arnepalli, V.B. Maji, Modelling carbon dioxide adsorption behaviour on montmorillonite at supercritical temperatures. Adsorption **30**(7), 1703–1716 (2024). https://doi.org/10.1007/s10450-024-00525-z

G. Rother, E.G. Krukowski, D. Wallacher, N. Grimm, R.J. Bodnar, D.R. Cole, Pore size effects on the sorption of supercritical CO2 in mesoporous CPG-10 silica. J. Phys. Chem. C **116**(1), 917–922 (2012). https://doi.org/10.1021/jp209341q

N. Siemons, A. Busch, Measurement and interpretation of supercritical $CO_2$ sorption on various coals. Int. J. Coal Geol. **69**(4), 229–242 (2007). https://doi.org/10.1016/j.coal.2006.06.004

# Chapter 6
# Reduction of Overpressure Due to Adsorption

As discussed earlier for the pressure space, $CO_2$ storage capacity and the overpressure are interrelated. The relation between $CO_2$ adsorption and the overpressure will be discussed in more detail in this chapter.

## 6.1 Effect of Phase Change on $CO_2$ Pressure

Supercritical $CO_2$ cannot be compressed to liquid. However, it can be adsorbed to form a denser phase due to attraction forces near solid surface. Although the free-phase $CO_2$ can be compressed to reach the density of adsorbed $CO_2$, much higher pressure is required. This is illustrated in Fig. 6.1, which shows that to reach the density of adsorbed $CO_2$, the pressure of the free-phase $CO_2$ needs to exceed 60 MPa at 40 °C.

According to the Polanyi model, the density of the adsorbed phase is not dependent on the gas pressure. Therefore, under lower pressures the density of the adsorbed phase can be substantially higher than the density of the free-gas phase. However, the thickness of the adsorbed layer depends on the free-gas pressure and the attraction forces of the solid surface. Thus, the adsorbed amount depends on the pressure and the specific surface. Higher pressure, larger specific surface area and stronger attraction forces from the surface will increase $CO_2$ adsorption on the pore surface. As a result, the effective density of $CO_2$ in the pore space will increase, enabling greater retention of $CO_2$ and reduction of the overpressure compared to the situation where $CO_2$ is stored as the fluid phase only.



J. Wang et al., *Carbon Dioxide Adsorption in Rock and Geological Storage of Carbon*, SpringerBriefs in Applied Sciences and Technology, https://doi.org/10.1007/978-3-031-90218-5_6

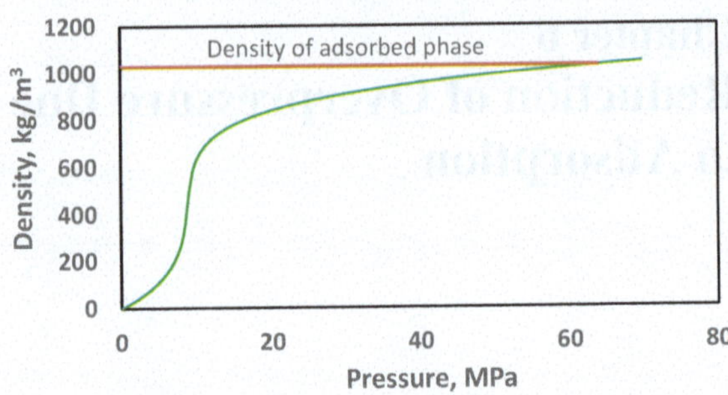

**Fig. 6.1** Pressure dependences of the densities of adsorbed $CO_2$ and free-phase $CO_2$



## 6.2  Pressure-Driven Phase Transition

For $CO_2$ storage, $CO_2$ must be injected into pore space at higher pressure than the pressure of the resident water. Without adsorption, the overpressure will be high. Adsorption by the pore surface leads to the phase transition from free-phase $CO_2$ to adsorbed $CO_2$ which has a higher density. As a result, the overpressure will decrease. This can be understood from the requirement for Gibbs free energy to decrease:

$$dG = -S dT + V dp \leq 0. \tag{6.1}$$

The term $S dT$ in Eq. (6.1) becomes zero under constant temperature. $V$ in Eq. (6.1) corresponds to the volume of $CO_2$ which equals the pore volume $V_p$. Accordingly,

$$dG = V_p dp \leq 0. \tag{6.2}$$

That is, the pressure $p$ will tend to decrease. Transition of free-phase $CO_2$ to adsorbed $CO_2$ is favored thermodynamically because it will decrease the density of the free-phase $CO_2$ by giving it more space, thereby resulting in lower pressure. Accordingly, the following relations will hold:

$$n_a = \rho_{am} V_a \tag{6.3}$$

$$n_f = \rho_{fm} V_f \tag{6.4}$$

$$n_a + n_f = n_{f0}, \tag{6.5}$$

where $n_a$, $n_f$ and $n_{f0}$ denote the number of moles for adsorbed $CO_2$, free-phase $CO_2$ and the free-phase $CO_2$ before the adsorption takes place, respectively. $\rho_{am}$ and $\rho_{fm}$ denote the molar density of adsorbed $CO_2$ and free-phase $CO_2$, respectively. Differentiation of the above equations yields

$$dn_a = \rho_{am} dV_a + V_a d\rho_{am}$$

$$= \rho_{am} dV_a (d\rho_{am} = 0 \text{ because } \rho_a \text{ is constant}) \tag{6.6}$$

$$dn_f = \rho_{fm} dV_f + V_f d\rho_{fm} \tag{6.7}$$

$$dn_a + dn_f = \rho_{am} dV_a + \rho_{fm} dV_f + V_f d\rho_{fm} = dn_{f0} = 0. \tag{6.8}$$

Noting that $dV_a = -dV_f$ in given pore space, one can get from Eq. (6.8)

$$-(\rho_{am} - \rho_{fm}) dV_f + V_f d\rho_{fm} = 0. \tag{6.9}$$

That is

$$\frac{d\rho_{fm}}{\rho_{am} - \rho_{fm}} = \frac{dV_f}{V_f}. \tag{6.9'}$$

Consequently,

$$-d \ln(\rho_{am} - \rho_{fm}) = d \ln V_f \tag{6.10}$$

and

$$\ln \frac{\rho_{am} - \rho_{fm}}{\rho_{am} - \rho_{fm0}} = \ln \frac{V_{f0}}{V_f}, \tag{6.11}$$

where $\rho_{fm0}$ is the initial molar density of the free-phase $CO_2$ in the pore space. Thus,

$$\frac{\rho_{am} - \rho_{fm}}{\rho_{am} - \rho_{fm0}} = \frac{V_{f0}}{V_f}. \tag{6.12}$$

Thus

$$\rho_{am} - \rho_{fm} = \frac{V_{f0}}{V_f}(\rho_{am} - \rho_{fm0}) \tag{6.13}$$

Because $\rho_{fm0} > \rho_{fm}$ so $\frac{\rho_{am} - \rho_{fm}}{\rho_{am} - \rho_{fm0}} > 1$, we have $V_f < V_{f0}$. That is, the volume of the free-phase $CO_2$ decreases following the adsorption. Substitute $V_{f0}$ with $V_p$ and $V_f$ with $(V_p - V_a)$ and further arrange the equation yields

$$\rho_{fm0} - \rho_{fm} = \frac{V_a}{V_p}(\rho_{am} - \rho_{fm}) > 0. \tag{6.14}$$

Accordingly,

$$\rho_{fm} < \rho_{fm0}. \tag{6.15}$$

That is, the density of the free-phase $CO_2$ in the pore space decreases due to the adsorption. Because the pressure of free-phase $CO_2$ decreases with decreasing density, the pore pressure decreases following $CO_2$ adsorption. The extent of the pressure decrease can be determined from known pressure dependence of $CO_2$ density (as shown in Fig. 6.1, for example), or by calculation using equations of states. For example, using the van der Waals equation of state the relation between the pressure and density is given by

$$p = \frac{RT}{V_m - b} - \frac{a}{V_m^2} = \frac{\rho_{fm}RT}{1 - b\rho_{fm}} - a\rho_{fm}^2. \tag{5.13'}$$

The pore pressure reduction due to $CO_2$ adsorption can thus be calculated to get

$$\Delta p = p(\rho_{fm0}) - p(\rho_{fm}). \tag{6.16}$$

## 6.3   The Effects of $CO_2$ Dissolution and $CO_2$ Adsorption on Overpressure Reduction

Dissolution of $CO_2$ in reservoir water will decrease the overpressure caused by $CO_2$ injection. However, the dissolution can take hundreds of years due to limitations in the mass transfer area and diffusion of $CO_2$ molecules through the aqueous phase along the pores. By contrast, adsorption is much faster and enables nearly instant reduction of the overpressure. This can be seen from Fig. 6.2, which shows that $CO_2$ adsorption in rocks completes in just over an hour.

The molar volume of $CO_2$ in brine is reported to be 35 L/kmol (Garcia 2001). In comparison, the molar volumes of adsorbed $CO_2$ and free-phase $CO_2$ are 42.7 L/kmol (as estimated from the van der Waals constant $b$) and 69.6 L/kmol (at 10 MPa and 40 °C), respectively. Injection of $CO_2$ into reservoirs largely increases the overpressure. Dissolution of free-phase $CO_2$ would reduce the overpressure but due to the slow process large overpressure would remain in the reservoirs for very long time. The dissolution requires dissolved $CO_2$ to move away from the $CO_2$-brine interface to reach non-saturated zones. The rate can be enhanced by convective mixing of $CO_2$-saturated brine and the surrounding unsaturated brine (Xu et al. 2006). The $CO_2$-saturated brine has a higher density compared to the unsaturated brine. The difference in density between the $CO_2$-saturated brine and the less dense surrounding brine creates a convective force, causing the denser solution to settle down toward the bottom. Meanwhile, more $CO_2$ can be dissolved by the unsaturated formation brine which moves in due to the convection. However, the convective mixing process is limited by narrow pores. In high-permeability rocks, the water can flow more easily, enhancing the mixing. In low-permeability reservoirs or zones, the convective mixing will be restricted. In this situation, the transfer of dissolved $CO_2$ away from the $CO_2$/

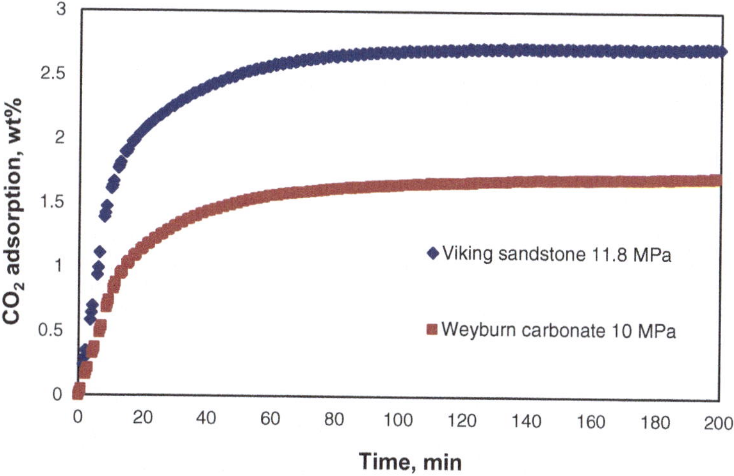

**Fig. 6.2** $CO_2$ adsorption on rock samples at 40 °C as a function of time



water interface depends on the slow diffusion of $CO_2$ molecules. Another point is that dissolution of $CO_2$ requires large quantities of water. For instance, based on reported data (Yan et al. 2011; Wang et al. 2019) for $CO_2$ solubility in brine (0.7 mol/kg at 10 MPa and 40 °C) and density of $CO_2$-brine solutions (about 1050 kg/m$^3$), dissolution of 1 volume of $CO_2$ would require about 20 volumes of brine. With limited contact with brine and limited convective mixing, dissolution of $CO_2$ will not be achievable in a short time frame. It has been estimated that the slow process only enables a 0.2–0.5% reduction per year for 10% overpressure caused by $CO_2$ injection (Peters et al. 2015). In contrast, adsorption takes place at the immediate contact with rock without the need of diffusion and is driven by pressure which is constantly maintained in the $CO_2$ phase. Unlike dissolution, adsorption does not require additional pore space to decrease the overpressure of injected $CO_2$. As has been shown earlier for the Weyburn rock, adsorption can reduce the overpressure by 80% in an hour (see Figs. 4.2 and 6.2). This is many orders of magnitude faster compared to dissolution. Whereas dissolution enables long-term trapping of $CO_2$ and greater pressure depletion, adsorption can be very effective for reducing the overpressure in a short timescale to bridge the gap until dissolution reaches a significant extent. Adsorbed $CO_2$ can also facilitate dissolution by increasing the contact area with water. This will be discussed later.

# References

J.E. Garcia, in *Density of Aqueous Solutions of $CO_2$*. U.S. Department of Energy Office of Scientific and Technical Information technical report 2001 (2001). https://www.osti.gov/biblio/790022. Accessed 23 Feb 2025

E. Peters, P.J.P. Egberts, D. Loeve, C. Hofstee, $CO_2$ dissolution and its impact on reservoir pressure behavior. Int. J. Greenhouse Gas Control **43**, 115–123 (2015). https://doi.org/10.1016/j.ijggc.2015.10.016

J. Wang, B. He, L. Xie, K. Bei, G. Li, Z. Chen, I. Chou, C. Lin, Z. Pan, Determination of $CO_2$ solubility in water and NaCl solutions under geological sequestration conditions using a fused silica capillary cell with in situ Raman spectroscopy. J. Chem. Eng. Data **64**(6), 2484–2496 (2019). https://doi.org/10.1021/acs.jced.9b00013

X. Xu, S. Chen, D. Zhang, Convective stability analysis of the long-term storage of carbon dioxide in deep saline aquifers. Adv. Water Resour. **29**(3), 397–407 (2006). https://doi.org/10.1007/s11432-006-0397-z

W. Yan, S. Huang, E.H. Stenby, Measurement and modeling of $CO_2$ solubility in NaCl brine and $CO_2$-saturated NaCl brine density. Int. J. Greenhouse Gas Control **5**(6), 1460–1477 (2011). https://doi.org/10.1016/j.ijggc.2011.08.004

# Chapter 7
# CO$_2$ Adsorption and Rock Wettability

As has been discussed above, adsorption of CO$_2$ can increase CO$_2$ uptake in pore space and decrease the pressure in the CO$_2$ fluid. These are highly desirable for CO$_2$ storage. However, for the adsorption to take place, CO$_2$ must displace the pore water and contact the pore walls. This can only be attained in a portion of the pores because rocks are water wet under normal conditions and the resultant capillary entry pressure, which is the pressure that a nonwetting fluid must overcome to enter the pores occupied by wetting fluids, opposes the entry of CO$_2$. The capillary entry pressure increases with decreasing pore size. As a result, injected CO$_2$ will bypass narrow pores or pore throats where the capillary effect is stronger, leaving a large fraction of the total pore surface unaccessed by CO$_2$.

## 7.1  Capillary Pressure and Wetting

Capillary pressure occurs when two immiscible fluids contact each other in narrow spaces of solids such as capillary tubes and narrow pore network of rocks. The pressure in the two fluids will be different and the pressure difference across the interface between the two fluid phases is called capillary pressure. The capillary pressure arises from capillary forces, which are interactions between fluid molecules and the solid surfaces surrounding the fluids. These forces include cohesive forces between like molecules and adhesive forces between fluid molecules and the solid surface. When the adhesive forces are stronger than cohesive forces of molecules of one fluid phase, the fluid is attracted by the solid and tends to spread over the solid surface. Such a fluid is regarded as the wetting fluid. When cohesive forces exceed adhesive forces, the fluid tends to contract on the solid surface and is regarded nonwetting. As a result, the wetting phase pushes the nonwetting phase over the solid surface until an equilibrium state is reached. A well-known phenomenon related to capillary pressure is capillary rise, where a wetting fluid displaces a nonwetting fluid



J. Wang et al., *Carbon Dioxide Adsorption in Rock and Geological Storage of Carbon*,
SpringerBriefs in Applied Sciences and Technology,
https://doi.org/10.1007/978-3-031-90218-5_7

in a capillary tube spontaneously. For example, when a glass capillary tube is dipped in water, the water inside the tube will rise to a height above the bulk water level. This behavior can be attributed to the adhesive forces between water and the glass surface, which are stronger than the cohesive forces between water molecules. As a result, water will move up to wet the wall and displace air in the tube. The capillary rise can be described by

$$p_c = p_{nw} - p_w = (\rho_w - \rho_{nw})gh, \tag{7.1}$$

where $p_c$ is the capillary pressure; $p_{nw}$ and $p_w$ are the pressure of the nonwetting phase and wetting phase, respectively; $\rho_w$ and $\rho_{nw}$ are the density of the wetting phase and nonwetting phase, respectively. $h$ is the height of the rise, i.e., the height of the column formed by the wetting phase inside the tube. In the case of water and air in the glass tube water is the wetting fluid and air is the nonwetting fluid. If the glass capillary tube is dipped in mercury, the cohesive forces of the mercury molecules are stronger than the adhesive forces with the glass surface and mercury inside the tube will be pushed down by air. In this situation air is the wetting fluid which wets the glass surface and mercury is the nonwetting fluid. Wetting and nonwetting are thus relative to specific fluid pairs and solid surface. A nonwetting fluid in one situation may become a wetting fluid in another situation. For instance, air in the glass capillary tube dipped in water is the nonwetting phase but becomes the wetting phase when the glass capillary is dipped in mercury. On a hydrophilic (water attracting) surface, water will be the wetting phase and the other fluid will be the nonwetting phase. However, on a hydrophobic (water repelling) surface water will be the nonwetting phase and the other fluid will be the wetting phase. The strength of capillary pressure can be measured by the height of the wetting-fluid column. Higher capillary pressure yields higher column.

A closer examination of the interface of the two fluids in the capillary tube will reveal that the interface is not flat but curved. The interface of the water and air is concave (curving down) and the interface of air and mercury is convex (curving up), as illustrated in Fig. 7.1. In general, the interface between two immiscible fluids in a narrow tube will always curve inward on the wetting phase side. The angle formed by the interface and solid surface which is measured through the wetting phase is called contact angle. An illustration of the contact angle is shown in Fig. 7.2. The contact angle reflects the relative wetness of the two contacting fluids and controls the capillary pressure. The nonwetting phase, which is always on the convex side, has higher pressure. This is described by the following formula:

$$p_c = p_{nw} - p_w = \frac{2\sigma \cos\theta}{r_c}, \tag{7.2}$$

where $\sigma$ is the interfacial tension between the two fluid phases, which is a force in all directions along the interface per unit length and tends to contract the interface. $r_c$ is the radius of the capillary. $\theta$ is the contact angle. The capillary pressure is thus related to the curved interface, which creates a net pressure on the nonwetting phase. This

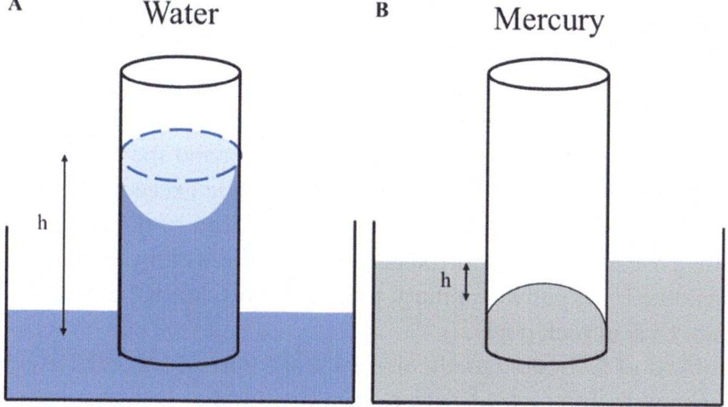

**Fig. 7.1** Illustrations of capillary actions. **a** Water and air; **b** mercury and air

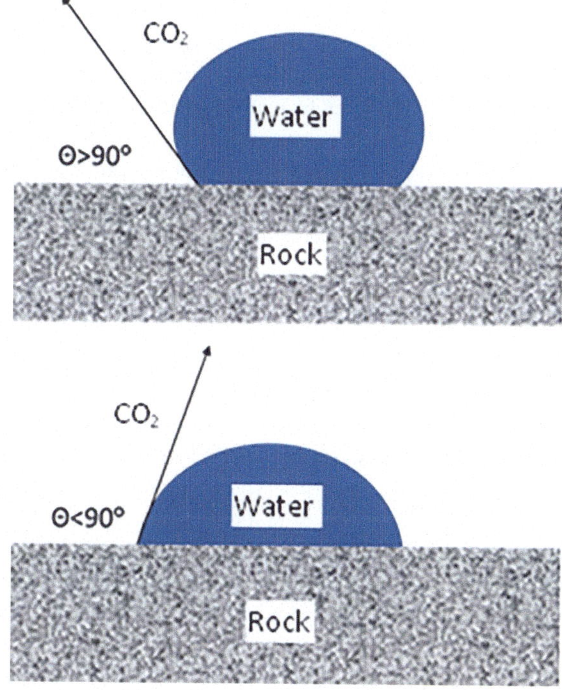


**Fig. 7.2** Illustration of the contact angle of a water drop

situation is not limited to capillary tubes. More generally, capillary pressure exists in any narrow space surrounded by solid surfaces where two immiscible fluid phases are in contact. The fluid which is attracted more by the solid surface is the wetting fluid and the fluid attracted less by the surface is the nonwetting fluid. Another point to note is that the capillary pressure does not only work vertically but works in all directions. For instance, in a narrow horizontal tube or pore originally filled with a wetting fluid, an additional pressure with the magnitude of the capillary pressure is required to force a nonwetting fluid into the narrow space. Conversely, if the narrow space is originally filled with a nonwetting fluid, a wetting fluid can enter the space spontaneously under a lower pressure than that of the nonwetting fluid. If the contact angle equals 90°, the capillary pressure becomes zero and the pressure in the two immiscible phases will be the same. If the contact angle exceeds 90°, however, the originally wetting phase will become nonwetting and can be displaced by the other phase. In this case the pressure difference between the wetting and the nonwetting fluid is considered a negative capillary pressure. This can happen to water under high-pressure $CO_2$ in rock pores.

In the context of $CO_2$ storage, the porous rocks may be modeled as a bundle of capillary tubes with different tube sizes. $r_c$ in the formula is then corresponding to the effective radius of rock pores. The smaller the radius, the larger the capillary pressure. This is because a smaller radius corresponds to a larger ratio of the contact area between the solid wall and the fluid column, so the wall's attraction effect on the wetting-fluid column is stronger. In the case of reservoir rocks, water is normally the wetting fluid which occupies the pore space of the rocks. $CO_2$ that is injected into the reservoirs would be the nonwetting fluid and need to be under higher pressure than the water in order to overcome the capillary effect and flow. As mentioned earlier, the capillary pressure increases with decreasing pore size, in water-wet rocks $CO_2$ would bypass the narrow pores or pore throats which oppose $CO_2$ entry by the capillary forces. Consequently, $CO_2$ occupancy of the reservoir pore space is low and the $CO_2$ plume moves to the top faster. This would increase the risk of $CO_2$ leakage and decrease reservoir utilization efficiency.

If the capillary pressure is lowered, $CO_2$ could enter the narrower pore space more easily. This not only increases $CO_2$ occupancy of the pore space but also increases $CO_2$ adsorption on the pore walls. As has been shown earlier, the capillary pressure is related to the pore radius, interfacial tension and contact angle. Whereas the radius does not change for given pore size, the interfacial tension and contact angle can change with process conditions for $CO_2$ injection. The interfacial tension between $CO_2$ and water has been observed to decrease with $CO_2$ pressure in experiments. The contact angle has been observed to increase with $CO_2$ pressure. The observations suggest that the product $\sigma \cos \theta$ could decrease with increasing $CO_2$ pressure. As has been mentioned earlier, the contact angle reflects the wettability of the solid surface. If the pore surface of the rock becomes less water-wet, the capillary pressure will be lower. Decreased water wettability of rocks under high-pressure $CO_2$, as indicated by increased contact angle, has been observed with various measurement methods (Sutjiadi-Sia et al. 2008; Yang et al. 2008; Kim et al. 2012; Jung and Wan 2012; Saraji et al. 2013; Iglauer et al. 2015; Iglauer 2018; Pan et al. 2018; Mutailipu et al. 2019; Fauziah et al. 2020) This is believed to be true for all rocks (Iglauer 2018). The rocks studied by Wang et al. (2023) for the present work also showed increased water contact angle with increasing $CO_2$ pressure, indicating increased $CO_2$ wetness.

As has been discussed above, $CO_2$ adsorption allows more $CO_2$ to be stored in the zones that $CO_2$ has reached. However, $CO_2$ can only reach a limited portion of reservoirs when the reservoir rocks are water wet. In this situation, the capillary force prevents injected $CO_2$ from entering small pores of rocks and lead to low utilization efficiency of the reservoirs. On the other hand, results of contact angle measurement suggest that rocks could become less water wet, implying that the capillary force could decrease under high-pressure $CO_2$. There are different suggestions over the mechanism for the wettability change, mostly based on the changes in the water phase due to dissolved $CO_2$. Based on the observations by Wang et al. (2023), however, the decreased water wetness can be related to $CO_2$ adsorption on the rocks, which can have a range of consequences for $CO_2$ storage, including better utilization of reservoir capacity, reduction of overpressure and immobilization of $CO_2$ fluid.

## 7.2 The Relation of Wettability Change to $CO_2$ Adsorption

The above equations for capillary force do not reflect the mechanism for capillary action. The driving force for the action can be understood from the Young's equation:

$$\cos\theta = \frac{\sigma_{sg} - \sigma_{sl}}{\sigma_{gl}}, \tag{7.3}$$

where $\theta$ is the contact angle of the wetting phase. $\sigma_{sg}$, $\sigma_{sl}$ and $\sigma_{gl}$ are the interfacial tension between solid surface and gas phase, solid surface and liquid phase, and gas phase and liquid phase, respectively. Each interfacial tension tends to contract the corresponding interface. The contact angle results from the balance of the contracting forces. The contact angle, which is the angle formed by a liquid at the boundary line with a gas or another liquid on the solid surface, is commonly regarded as a measurement of the wettability of the surface by fluids. The angle reflects the interactions of the solid surface and the contacting fluid phases, which control the wettability. The simplest and most common method to measure contact angle is the sessile drop method, where a drop of the liquid is placed on the surface of the solid surrounded by the other fluid. Examples are shown in Fig. 7.3. When the density of the liquid drop is lower than the density of the other fluid, the buoyancy force will move the liquid drop upward and the sessile drop method cannot be successful. In this case a captive bubble method, which lets the lighter fluid phase to form a bubble on the bottom surface of the solid surface submerged in the heavier fluid phase, can be used. In both methods, the wetting phase will have a contact angle smaller than $90°$ and tend to spread over or adhere to the solid surface. The nonwetting phase will have an angle larger than $90°$ and not tend to adhere to the surface. The two angles measured from the two fluid phases add up to $180°$. The interfacial tension between the two fluid phases $\sigma_{gl}$ can be measured as a function of pressure, by pendant drop method (Wang et al. 2019), for example, at given temperatures. By contrast, the other two interfacial tensions $\sigma_{sg}$ and $\sigma_{sl}$ cannot be measured separately. One can only determine the difference of the two interfacial tensions from the Young's equation:

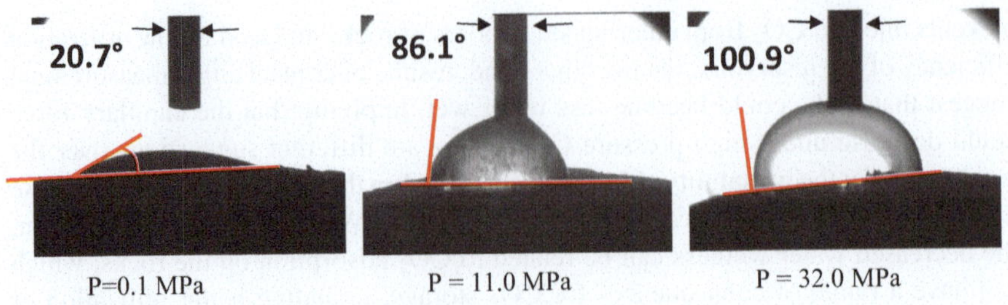


**Fig. 7.3** Reservoir water on the surface of Viking sandstone sample in $CO_2$ under different pressures at 40 °C (from Wang et al. 2019)

$$\sigma_{sg} - \sigma_{sl} = \sigma_{gl}\cos\theta \tag{7.3'}$$

It can be seen from Eq. (7.3′) that if $\sigma_{sg} - \sigma_{sl}$ does not change but $\sigma_{gl}$ decreases, $\cos\theta$ must increase. Normally the interfacial tension between a gas phase and a solid surface will be larger than the interfacial tension between a liquid phase and the solid surface, i.e., $\sigma_{sg} > \sigma_{sl}$. Hence, the interface between the gas and the solid surface will tend to decrease and $cos\theta$ will have a positive value: $0 < \cos\theta \le 1$, which corresponds to $\theta < 90°$. In this state the solid surface is said to be wet by the liquid. If, however, the solid surface becomes gas wet, there will be $\sigma_{sg} < \sigma_{sl}$ and $\cos\theta < 0$ according to Eq. (7.3′). This results in $\theta > 90°$ if the contact angle is measured from the liquid phase.

In the experiments by Wang et al. (2022, 2023), it has been observed that $\sigma$ between $CO_2$ and water decreases with increasing $CO_2$ pressure whereas $\theta$, the water contact angle increases with $CO_2$ pressure on several rocks, as shown in Figs. 7.4 and 7.5. Similar observations have been reported by other researchers. In consequence, $\sigma\cos\theta$ decreases with increasing $CO_2$ pressure, suggesting decreasing difference between $\sigma_{sg}$ and $\sigma_{sl}$ with increasing $CO_2$ pressure. Various views by other researchers have been proposed regarding the increased water contact angle by $CO_2$, including change of electric double layer on water/rock interface and change of zeta potential caused by dissolved $CO_2$. However, a correlation between the contact angle change and adsorption of $CO_2$ on rocks has been noticed by Wang et al. (2023), i.e., both the contact angle in $CO_2$ atmosphere and $CO_2$ adsorption increases with increasing $CO_2$ pressure (Fig. 7.5).

As discussed earlier, the contact angle can be expressed in terms of three interfacial tensions: $\sigma_{sg} - \sigma_{sl} = \sigma_{gl}\cos\theta$. As $\sigma_{gl}$ is not dependent on rock surface, the observed change of the contact angle on rock surface can be attributed to the change in $\sigma_{sg} - \sigma_{sl}$ with increasing $CO_2$ pressure. It is worth noting that, although $\sigma_{gl}$ does not affect the rock surface, it affects the contact angle. When $\sigma_{sg} - \sigma_{sl}$ does not change, the contact angle will decrease with decreasing $\sigma_{gl}$ according to Young's equation. As $\sigma_{gl}$ is observed to decrease with increasing $CO_2$ pressure but the contact angle increases with $CO_2$ pressure, there must be changes in $\sigma_{sg} - \sigma_{sl}$ with increasing $CO_2$ pressure.

The difficulty with applying Eq. (7.3′) is that $\sigma_{sg}$ and $\sigma_{sl}$ cannot be measured directly. Besides, $\sigma_{gl}$ changes with pressure, as shown by the results from pendant

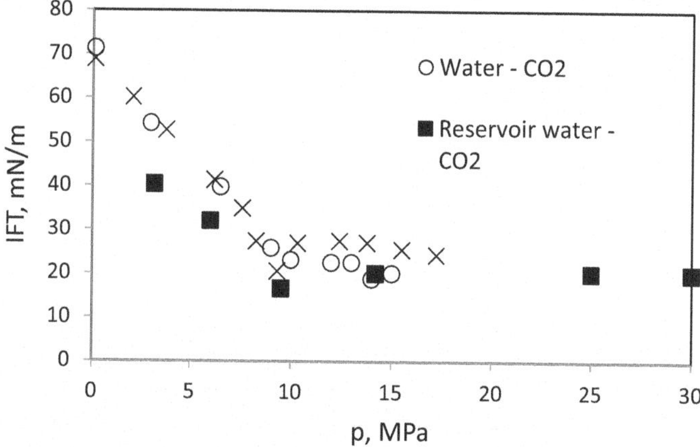



**Fig. 7.4** Interfacial tension between water and CO$_2$ as a function of CO$_2$ pressure at 40 °C (adapted from Wang et al. 2023). The reservoir water was extracted from the Viking sandstone by supercritical CO$_2$ (Wang et al. 2019). Literature data (represented by the x symbol) were obtained using a capillary rise method (Chun and Wilkinson 1995)

**Fig. 7.5** Pressure dependence of contact angle and excess adsorption of CO$_2$ on rocks at 40 °C (data from Wang et al. 2023)

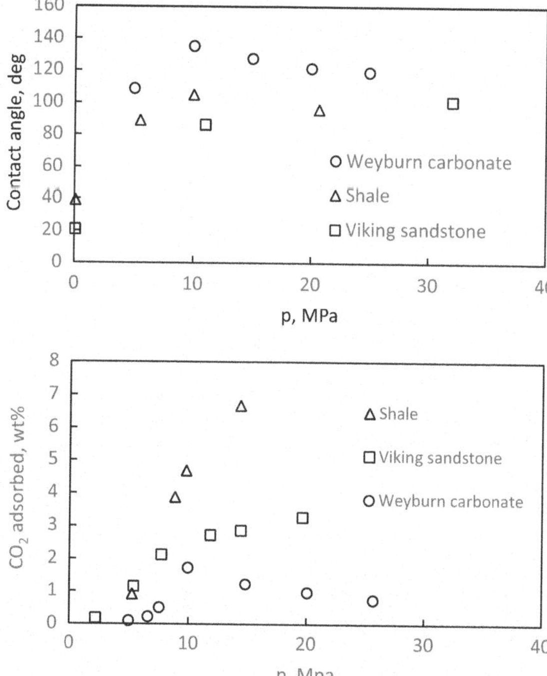

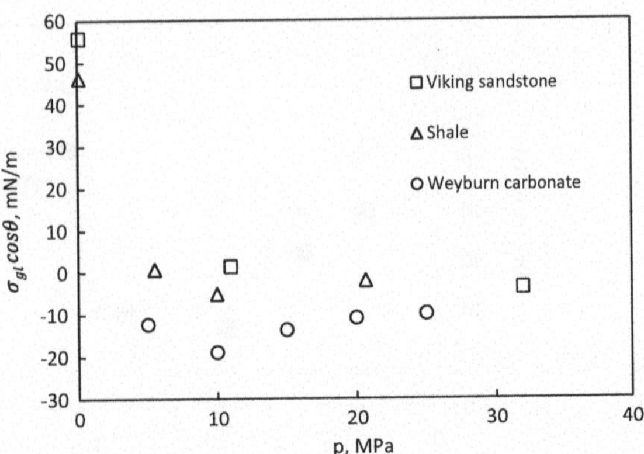



**Fig. 7.6** Dependence of $\sigma_{gl}\cos\theta$ on CO$_2$ pressure at 40 °C (from Wang et al. 2023)

drop experiments in Fig. 7.4 in comparison with $\sigma_{gl}$ values reported by others using a capillary rise method. In the present work separately measured values of $\sigma_{gl}$ and $\theta$ under varied CO$_2$ pressures have been used to determine the values of $\sigma_{gl}\cos\theta$, and the $\sigma_{gl}\cos\theta$ values are plotted against CO$_2$ pressure (Fig. 7.6) to analyze the pressure dependence of $\sigma_{sg} - \sigma_{sl}$ and explore possible causes to the change of the contact angle. However, $\sigma_{sg}$ and $\sigma_{sl}$ are not intuitive properties for analyzing the effect of pressure. To enable the analysis, $\sigma_{sg}$ is related to a variable which reflects the effect of adsorbed molecules on the surface:

$$\sigma_{sg} = \sigma_s^0 - \pi_g, \tag{7.4}$$

where $\pi_g$ is termed spreading pressure, which acts in the opposite direction of surface tension on the two-dimensional interface. $\sigma_s^0$ is the surface energy of the solid prior to contacting a fluid phase. For the present analysis $\sigma_{sl}$ is further expressed as

$$\sigma_{sl} = \sigma_s^0 - \pi_l, \tag{7.5}$$

where $\pi_l$ represents the spreading pressure on the water side. With the two new variables the Young's equation can be written into

$$\sigma_{gl}\cos\theta = \pi_l - \pi_g \tag{7.6}$$

The difference between Eqs. (7.6) and (7.3′) is that the spreading pressure tends to expand, rather than decrease, the interface with solid surface. The contact angle in Eq. (7.6) can be viewed as the result of the competition of two opposing spreading pressures pushing the contact line. By plotting $\sigma_{gl}\cos\theta$ against CO$_2$ pressure, one can see clearly in Fig. 7.6 that $\pi_l - \pi_g$ (i.e., $\sigma_{gl}\cos\theta$) decreases with CO$_2$ pressure, at least in the pressure range up to 10 MPa. By neglecting the dependence of $\pi_l$ on CO$_2$ pressure in this range, the change of $\pi_g$, the spreading pressure on the CO$_2$ side of the contact line can be related to adsorption on the rock from the CO$_2$ phase.

The Gibbs equation for the surface tension at the solid–gas interface can be written as (Hiemenz 1986; Condon 2019; Butt et al. 2006):

$$d\sigma = -\Gamma \, d\mu_a \tag{7.7-1}$$

$$\Gamma = \frac{n_a}{A} \tag{7.7-2}$$

This equation establishes that adsorption of gas molecules on the solid surface decreases the surface free energy, where $\sigma$ is the surface energy; $\Gamma$ is the number of adsorbed molecules $n_a$ per unit surface area $A$; $\mu_a$ is the chemical potential of adsorbed molecules, which equals the chemical potential of the molecules in gas phase $\mu_g$ at equilibrium. Accordingly,

$$\mu_a = \mu_g = \mu_g^0 + RT \ln f \tag{7.8-1}$$

$$d\mu_a = d\mu_g = RT d \ln f, \tag{7.8-2}$$

where $\mu_g^0$ is the chemical potential at the standard state. $R$ is the gas constant. $T$ is the absolute temperature. $f$ is the fugacity of the gas molecules. According to the definition of fugacity

$$RT d \ln f = V_g dp, \tag{7.9}$$

where $V_g$ is the molar volume of the gas and $p$ is the gas pressure. Equation (7.7) can be written into

$$-d\sigma = \frac{n_a V_g}{A} dp = \frac{n_a V_g}{A_s M} dp, \tag{7.10}$$

where $A_s$ and $M$ are the specific surface area and mass of rock, respectively. The conversion of $A$ to $A_s M$ in Eq. (7.10) is for using adsorption data that is given as the number of adsorbed molecules per unit mass of the rock. Applying Eq. (7.4) to the surface tension $\sigma$ in Eq. (7.10), one obtains

$$\sigma_s^0 - \sigma_{sg} = \pi_g = \frac{1}{A_s} \int_0^p \frac{n_a V_g}{M} dp \tag{7.11}$$

Thus, the spreading pressure $\pi_g$ is related to gas pressure $p$. If the adsorption data can be described by an equation, for example the Langmuir equation shown in Chap. 2

$$C = \frac{C_m bp}{1 + bp}, \tag{2.3}$$

where $C$ is the amount of adsorbed molecules per unit mass of rock, and $C_m$ and $b$ are parameters independent of pressure, one can obtain from Eq. (7.11) and the Langmuir equation by noting that $\frac{n_a}{M} = C$ and therefore $\pi_g = \frac{1}{A_s} \int_0^p \frac{n_a V_g}{M} dp = \frac{1}{A_s} \int_0^p \frac{C_m b p V_g}{1+bp} dp = \frac{1}{A_s} \int_0^p \frac{C_m b Z R T}{1+bp} dp = \frac{\overline{Z} R T}{A_s} \int_0^p \frac{C_m b}{1+bp} dp$, where $Z = pV_g/RT$ is the compressibility factor and $\overline{Z}$ is the mean value of $Z$ over the range of integration according to the mean value theorem. In consequence,

$$\pi_g = \frac{\overline{Z} R T C_m}{A_s} \int_0^p \frac{b\,dp}{1+bp} = \frac{\overline{Z} R T C_m}{A_s} \int_0^p \frac{d(1+bp)}{1+bp} = \frac{\overline{Z} R T C_m}{A_s} \ln(1+bp) \quad (7.12)$$

However, the adsorption data in Fig. 7.5 cannot be described by the Langmuir equation and other commonly used equations. Besides, the volume of CO$_2$ depends strongly on the pressure, making the evaluation difficult. In this work, an approximate method is used to examine the magnitude of the spreading pressure by using the sum of the products $\left(\frac{n_a}{M}\right) V_g \Delta p$ to represent the integral in Eq. (7.11). Here the volume $V_g$ is determined from the density corresponding to CO$_2$ adsorption pressure and the integrand $\left(\frac{n_a}{M}\right) V_g$ represents the volume of the free-phase CO$_2$ which is adsorbed on unit mass of rock; $\Delta p$ is the increment of pressure from the adsorption run under lower CO$_2$ pressure. The method is illustrated in Fig. 7.7 for the Weyburn carbonate data, where the value of the integral is approximated by the area under the polyline that connects the data points. The area is evaluated using the trapezoidal rule and the associated error could be within several percent for the data. The results of the numerical approximation are shown in Fig. 7.8, which represent accumulated free energy of adsorbed CO$_2$ on unit mass of rock over the course of pressure increase. These values divided by the specific surface area of the rock represent the free energy on unit surface area of the rock on the CO$_2$ side. For given adsorption uptake, a smaller specific surface area corresponds to higher energy density which results in higher potential to push the contact line, leading to increased contact angle.

Now we compare the results shown in Fig. 7.9, which account for the contribution of CO$_2$ adsorption to the spreading pressure from the CO$_2$ side, with the results of contact angle and IFT measurements. Assuming negligible dependence of $\pi_l$ on CO$_2$ pressure in Eq. (7.6), we fit the results from adsorption to the results of contact angle and IFT. Figure 7.9 shows that the surface energy change calculated from the CO$_2$ adsorption data exceeds the change calculated from the contact angle and IFT data. To yield the pressure dependence of $\sigma_{gl} \cos\theta$ in lower CO$_2$ pressure range, the specific surface area of Weyburn rock needs to reach 20 m$^2$/g, which is 20 times higher than reported values for Weyburn and other carbonates (Wang et al. 2023). In other words, the surface energy change of Weyburn rock due to CO$_2$ adsorption is 20 times larger than is required to cause the contact angle change. In the case of Viking sandstone, the required specific surface area is 13 m$^2$/g, which is twice higher than reported values (see, e.g., Cao et al. 2016). It can also be seen that the contact angle of Weyburn rock changes in a smaller range, which is consistent with lower adsorption of CO$_2$ by this rock. In the case of shale, the required specific surface area

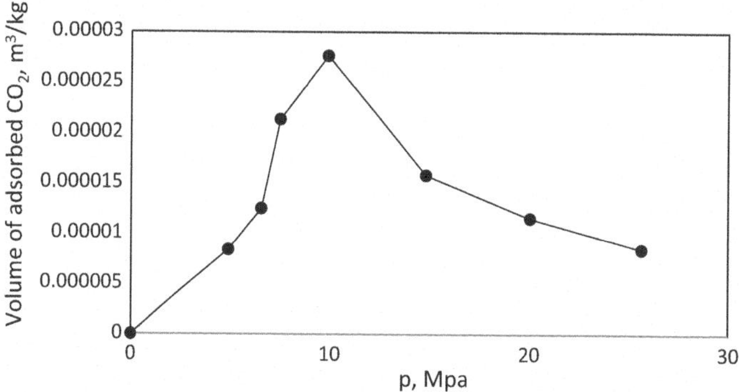

**Fig. 7.7** Numerical evaluation of the spreading pressure in terms of Eq. (7.11) for adsorption data of Weyburn carbonate. The solid symbol represents the integrand in Eq. (7.11), which is the product of the molar amount of adsorbed $CO_2$ and the molar volume at each pressure (Wang et al. 2023)

is 6 m$^2$/g, which is within the range of reported values. It is worth mentioning that adsorption of $CO_2$ on shale is known to depend on organic matter and clay content, which may not be two-dimensional and may not correlate well with specific surface area. Nevertheless, the change of the contact angle could correlate with adsorption of $CO_2$ on this shale. The results imply that the spreading pressure on the water side $\pi_1$ must increase under high $CO_2$ pressure and overweigh the increase of $\pi_g$, according to Eqs. (7.6) and (7.11). That is to say, the surface energy on the water side must decrease under high $CO_2$ pressure. It should be noted that the lowest value of $\sigma_{gl} \cos \theta$ is limited to about $-20$ mN/m by the lowest IFT (about 20 mN/m according to Fig. 7.4) and smallest $\cos \theta$ ($\cos 180° = -1$), so the calculated values below this limit are hypothetical. Above the $CO_2$ pressure that corresponds to the limiting value of $\sigma_{gl} \cos \theta$ the rock would be completely $CO_2$ wet, provided the spreading pressure on the side of water phase does not increase with $CO_2$ pressure. However, the present contact angle data suggest that the spreading pressure on the water side increases and results in a decrease of the contact angle (increase of $\sigma_{gl} \cos \theta$) at higher $CO_2$ pressure, as seen from the results of Weyburn carbonate and the shale in Fig. 7.5 above 10 MPa. One possible reason is that under high $CO_2$ pressure, the concentration of dissolved species in the water increases, which could adsorb to the water–rock interface and increase the spreading pressure on the water side. Change of surface charges of the rock/water and water/$CO_2$ interfaces could also contribute to the change of the contact angle (Kim et al. 2012; Saraji et al. 2013) The presence of ionic species in water can affect the wettability of rock surfaces, and the specific effect depends on various factors, including the type and concentration of ions, the mineral composition of the rock, and the existing wettability state of the rock. Some ionic species can increase the water-wetness of rock surfaces in oil fields. For example, divalent cations like calcium ($Ca^{2+}$) and magnesium ($Mg^{2+}$) can promote water wetting by reducing the adhesion of oil to the rock surface. Conversely, some ions can make rock surfaces less water-wet. For instance, high concentrations of

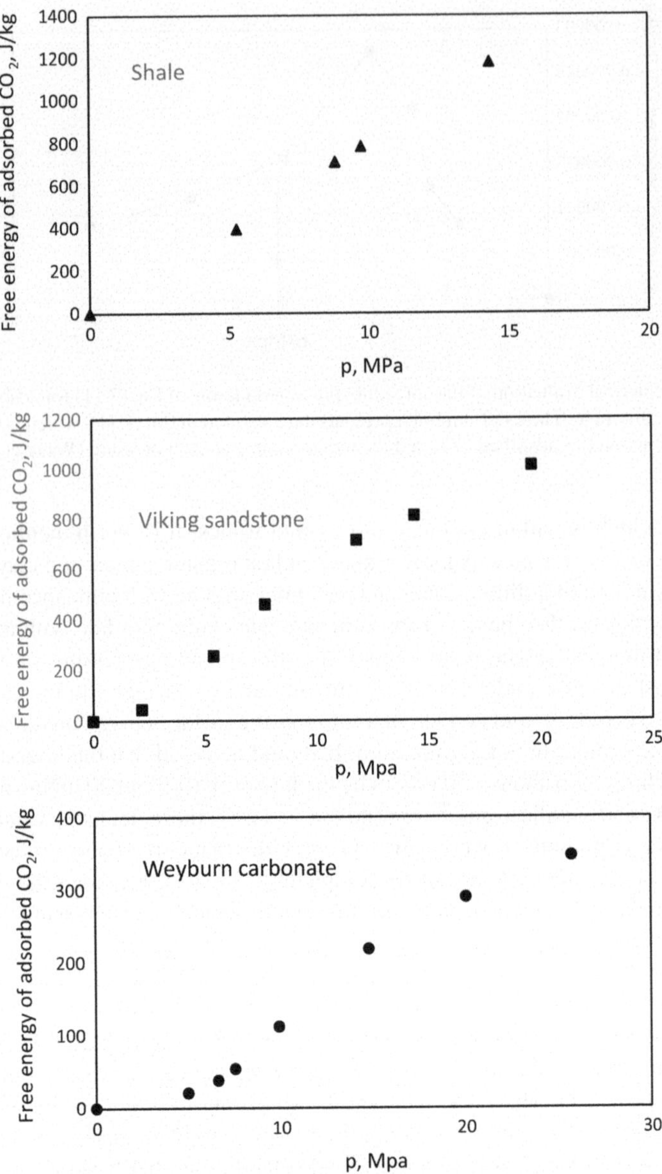



**Fig. 7.8** Free energy of adsorbed $CO_2$ on rock surface at 40 °C

certain monovalent cations like sodium (Na$^+$) can lead to the formation of a more oil-wet surface, especially if the rock surface has organic deposits or if the rock is initially oil-wet. Proposed mechanisms of wettability alteration by ionic species include ion exchange and adsorption, double layer expansion and surface complexation. In ion exchange and adsorption, ionic species can participate in ion exchange reactions with the mineral components of the rock. This can cause changes to the surface charge and change the wettability. In double layer expansion, presence of ionic species can affect the thickness of the electrical double layer at the rock-water interface. An expanded double layer due to higher ionic strength can lead to increased repulsion between oil and the rock surface, thereby enhancing water wetness. In surface complexation, ions form complexes with surface functional groups on the rock minerals, which alter surface chemistry and wettability. For example, divalent cations can bridge between negatively charged surface sites and polar oil molecules, thus reducing oil adhesion. Commonly considered ionic species for wettability alteration include calcium (Ca$^{2+}$), magnesium (Mg$^{2+}$) and sodium (Na$^+$). Ca$^{2+}$ and Mg$^{2+}$ are known to increase water-wetness by forming strong bonds with rock surface, displacing oil molecules, and promoting adsorption of water molecules. Na$^+$ in high concentrations can compete with Ca$^{2+}$ and Mg$^{2+}$ for adsorption sites. This may reduce water wetness. It is also known that adsorption of surfactants in crude oil and/or deposition of organic matter can alter wettability of rock surface (Anderson 1986). The experimental results shown above, however, suggest that adsorption of CO$_2$ can cause the wettability change as well. Wetting requires adhesion of the fluid phase to the solid surface. Adsorption depends on adhesive forces and can therefore be taken as a specific type of adhesion. In the present case, adhesion of the wetting phase can be realized by adsorption on the CO$_2$ side.

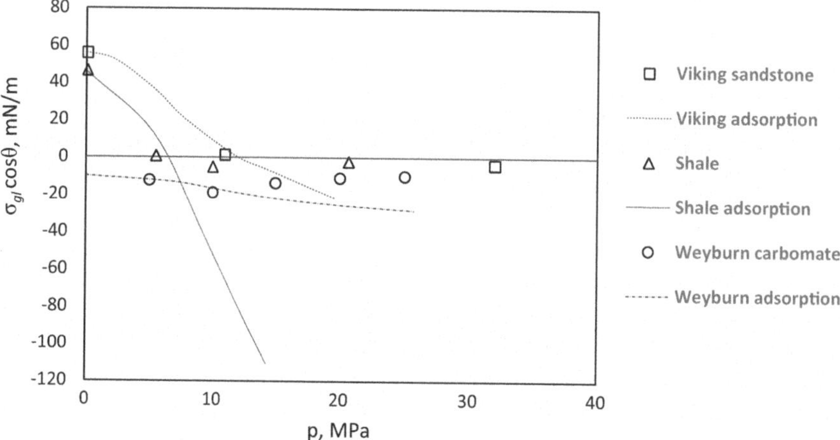


**Fig. 7.9** Comparison of the results of contact angle and IFT with the results of adsorption. The values from contact angle and IFT measurements are represented by symbols and the values from adsorption measurements are represented by the lines

| t (min) | 0 | 5 | 10 | 15 | 20 | 25 | 30 | 35 | 40 |
|---|---|---|---|---|---|---|---|---|---|
| **SW-DIW** | | | | | | | | | |
| V (mm$^3$) | 7.88 | 4.18 | 1.19 | 0.354 | 0.131 | 0.072 | 0.033 | 0.014 | 0.0025 |
| $\theta_{CO2}$ (°) | 136 | 122 | 96 | 92 | 91 | 85 | 81 | 59 | 25 |
| **DPL-DIW** | | | | | | | | | |
| V (mm$^3$) | 8.19 | 4.02 | 1.14 | 0.058 | 0.019 | 0.005 | 0 | | |
| $\theta_{CO2}$ (°) | 123 | 121 | 120 | 70 | 55 | 55 | 0 | | |
| **SW-SSW** | | | | | | | | | |
| V (mm$^3$) | 6.94 | 6.42 | 6.17 | 5.69 | 5.47 | 5.14 | 4.83 | 4.63 | 4.44 |
| $\theta_{CO2}$ (°) | 122 | 118 | 110 | 105 | 97 | 90 | 89 | 86 | 86 |
| **DPL-SSW** | | | | | | | | | |
| V (mm$^3$) | 8.03 | 7.37 | 7.14 | 6.67 | 6.13 | 5.83 | 5.61 | 5.45 | 5.13 |
| $\theta_{CO2}$ (°) | 139 | 139 | 138 | 137 | 137 | 135 | 136 | 136 | 136 |
| **SS-SSW** | | | | | | | | | |
| V(mm$^3$) | 8.94 | 8.18 | 7.76 | 7.30 | 7.10 | 6.78 | 6.46 | 6.28 | 6.10 |
| $\theta_{CO2}$ (°) | 138 | 137 | 138 | 137 | 138 | 138 | 137 | 135 | 135 |

**Fig. 7.10** CO$_2$ bubble size and contact angle variations as a function of time at 15 MPa and 40 °C (from Irfan et al. 2018). Note that decreasing CO$_2$ contact angle $\theta_{CO_2}$ translates to increasing water contact angle, indicating increased CO$_2$ wetness

Although simultaneous adsorption and wettability change is difficult to observe experimentally, it can be inferred from other observations. Irfan et al. (2018) reported interesting behaviors of CO$_2$ captive bubbles formed beneath two rock samples and a silica wafer submerged in water under 100 bar pressure. As shown by the images in Fig. 7.10, the volume of the bubbles and the contact angles of the bubbles on the rocks, one was a sandstone and another was a limestone, and a silica wafer changed with time. With the volume decreasing, the contact angle of CO$_2$ decreased, indicating the increase of the contact angle of water. These changes suggest that the surfaces become more CO$_2$ wet with time. The decreasing volumes of the bubbles on the surface of silica wafer, which is nonporous, could be attributed solely to dissolution into water. By contrast, the decreasing volumes on the surfaces of the porous rocks could be partly due to CO$_2$ entering the pores. However, the increasing water contact angles on all three materials suggest changed wettability of the surfaces toward more CO$_2$ wet. The changed wettability could be related to adsorption of CO$_2$ on the surfaces of the rocks and the silica wafer at the contact lines of the bubbles and water. The adsorption can reduce the water-wetness of the surfaces and decrease the capillary entry pressure and allow CO$_2$ to displace pore water held by capillary forces. As a result, the CO$_2$ phase entered the pores and the volumes of the bubbles decreased. It should be noted that the buoyant force of the CO$_2$ bubbles could also contribute to the entry of CO$_2$ into the pores, but this contribution is likely to be small because of the small size of the bubbles (the buoyant force would be proportional to the volumes of the bubbles). The time scale of the contact angles change, which occurred in 40 min, is also comparable to that of $CO_2$ adsorption (Fig. 6.2). From Fig. 7.10 it can be seen that the wetness change is a dynamic process which can be related to $CO_2$ adsorption process. The time dependence of the $CO_2$ wetness can be attributed to $CO_2$ adsorption on the surface, which increases with time.

## 7.3   The Tendency of $CO_2$ to Adsorb on Rock Pore Surface and Change the Wettability

### 7.3.1   *Higher Tendency for Supercritical Fluids to Adsorb on Solid Surfaces*

Under reservoir conditions, $CO_2$ is a supercritical fluid. Supercritical fluids exhibit several unique properties in common. Just above the critical temperature and critical pressure, supercritical fluids have densities between that of gases and liquids, and the densities are sensitive to temperature and pressure. In the case of $CO_2$, the density increases drastically around the critical pressure. By contrast, normal gases such as $H_2$, $CH_4$ and $N_2$ behave more like ideal gases whose density-pressure relationship is linear. As a result, their density increase is much smaller than that of $CO_2$ in the same pressure range. The higher density of $CO_2$ results in higher free energy per unit volume due to compression, leading to higher tendency of adsorption to reduce the surface energy of the pore walls as implied by the Gibbs equation. Moreover, as discussed earlier, the adsorbed layer is liquid-like (according to Polanyi's theorem), which has a higher density than that of the free phase. Higher density requires stronger attractions not only from the solid surface but also between the adsorbate molecules. The attraction forces increase with decreasing distances between the molecules. During pressurization, the adsorbate molecules in the free phase get closer and the intermolecular attractions increase, increasing the tendency for the free phase to further densify by transition to the adsorbed phase. This situation may be compared with vapor condensation, to which adsorption phenomenon has been analogized (see, e.g., Butt et al. 2006). When vapors are compressed, they condense to liquids. For supercritical $CO_2$ in the vicinity of the critical point, the fluid density increases rapidly with increasing pressure, creating a favorable condition for the transition to the higher density phase. By comparison, the normal gases cannot be densified enough to reach the condition for the phase transition even under very high pressure. This is consistent with the fact that the normal gases are less adsorbable than $CO_2$.

Another point for the higher degree of adsorption is that $CO_2$ is heavier than common nonpolar gases such as $H_2$, $CH_4$ and $N_2$. Heavier nonpolar molecules tend to be more apt to be adsorbed compared to lighter ones, and several factors can contribute to this phenomenon:

**Table 7.1** Free energy changes of $CO_2$ and two common gases due to pressurization from 1 to 100 bar at 40 °C

| Species | Density (kg/m$^3$) | Volume (m$^3$/kg) | Gibbs free energy change (MJ/m$^3$) |
|---|---|---|---|
| $CO_2$ | 629 | 0.0016 | 156 |
| $N_2$ | 106 | 0.009 | 46 |
| $CH_4$ | 70 | 0.014 | 52 |

1. Adsorption of nonpolar molecules is primarily driven by van der Waals forces, which include London dispersion forces. Heavier molecules usually have larger electron clouds, resulting in stronger dispersion forces due to increased polarizability. This makes heavier nonpolar species easier to be adsorbed.
2. Heavier molecules are often larger, which can lead to a greater surface area available for interaction with the adsorbent material. This increased contact area enhances the likelihood of adsorption.
3. Heavier molecules have lower velocity at given temperatures due to their greater mass, which can result in longer residence times at the surface, increasing the chances of adsorption.

Accordingly, heavier nonpolar molecules ($CO_2$ in the present case) would have a greater tendency to be adsorbed due to their size, polarizability, and the strength of van der Waals interactions.

In Table 7.1 we compare the free energy changes due to pressurization from 1 to 100 bar at 40 °C, which is relevant to $CO_2$ storage conditions, for $CO_2$ and two common gases $N_2$ and $CH_4$.

The results imply that $CO_2$ has much higher tendency to be adsorbed than the other two gases. The higher adsorption tendency of $CO_2$ has also been confirmed by quantum chemistry calculations (see, e.g., Jiang 2018).

### 7.3.2   *Effect of Free Energy Density on Wettability*

Because the free energy change owing to raised pressure for $CO_2$ is large, the surface energy of the rocks can be reduced. By contrast, the free energy change in water is small. The effect of pressure on the wettability can be significant because of the free energy change on the two sides of the contact line. Unless there are significant changes in the concentration of ionic species with increasing pressure, the free energy on the water side would not change as much as the free energy on the $CO_2$ side. Accordingly, the change of the wettability can be pressure dependent. For a quick estimate, we compare the free energy changes due to pressurization for $CO_2$ and water in Table 7.2. It can be seen that pressurizing water from atmospheric pressure (approximately 1 bar) to 100 bar causes little volume change and Gibbs free energy change. By contrast, the volume and the free energy of $CO_2$ change greatly after the same degree of pressurization. Although solutes in the water phase are not considered, the values

the fraction of adsorbed phase in the space taken by $CO_2$, i.e.,

$$\frac{M_a}{M_t} = \frac{M_{ex}}{\left(1 - \frac{\rho_f}{\rho_a}\right)M_t},$$

where $M_a$ is the actually adsorbed $CO_2$, which is viewed as immobile; $M_{ex}$ is the directly measured excess adsorption of $CO_2$ and $M_t$ is the total $CO_2$ which can be obtained from Eq. (4.7). To make a quick estimate of this fraction, we use again the value 1030 kg/m$^3$ for the density of the adsorbed phase. The resultant estimate for adsorbed $CO_2$ in the Viking rock is over 70% of the total $CO_2$ at 100 bar and 40 °C. This can be taken as the fraction of $CO_2$ that will be trapped.

Krevor et al. (2015) collected extensive literature data on residual trapping, which are shown in Fig. 9.3. The data for target $CO_2$ storage reservoirs in the Alberta Basin, Canada suggest that the trapping efficiency ranged from 20 to 60% (Krevor et al. 2015). Based on the above analysis, trapping by adsorbed $CO_2$ in the Viking rock sample exceeds the reported residual trapping level. It should be noted that the Viking rock sample is from a tight part of Viking reservoir, where the permeability is low and hydraulic fracturing has been used for oil and gas production. As has been discussed earlier, tight rocks generally have smaller pore size which favors $CO_2$ adsorption. The Viking rock sample indeed exhibits high $CO_2$ adsorption capacity, as shown in Fig. 2. 4. Rocks with lower $CO_2$ adsorption capacity may have lower trapping efficiency. In the case of the Weyburn reservoir discussed earlier, the trapping efficiency will become 46%. This value is still remarkable in comparison with the 20–60% trapping efficiency for the target $CO_2$ storage reservoirs evaluated by Krevor et al. (2015), and suggests that adsorbed $CO_2$ could make a significant contribution to $CO_2$ trapping even if the rock wettability is altered by $CO_2$.

Capillary pressure hysteresis and relative permeability hysteresis have been related to residual trapping in the literature (see, e.g., Burnside and Naylor 2014; Rasmusson et al. 2018). The present desorption results suggest that a different type of hysteresis—desorption hysteresis—could have a large impact on trapping of $CO_2$.

Thus, whereas wettability alteration induced by $CO_2$ may reduce the potential for residual $CO_2$ trapping via the snap-off mechanism, it can increase $CO_2$ adsorption. Since adsorbed $CO_2$ can stay immobilized after pressure decreases, trapping of $CO_2$ after water invasion is attainable. The trapping may not be achieved by snap-off but could be achieved due to desorption hysteresis. The increased $CO_2$ wetness of rocks would also allow more contact of water with the adsorbed $CO_2$ layer on pore walls which expedites $CO_2$ dissolution and increase trapping of $CO_2$ as dissolved species.

For comparison of the impact of residual trapping and adsorption trapping on $CO_2$ storage, more factors should be considered. Residual trapping is only important in the post-injection stage as it requires water invasion to zones where $CO_2$ pressure has declined. It has been argued in favor of residual trapping over mineral trapping that, whereas mineralization guarantees long-term trapping of $CO_2$, it takes effect after very long time. In contrast, residual trapping takes place over years to decades, which is within the operational period of CCS projects (Huber et al. 2016). This

**Fig. 9.3** Data for residual trapping of CO$_2$ collected by Krevor et al. (2015). The upper plot shows reported observations of residual trapping with supercritical CO$_2$ made on *Berea sandstones*. The lower graph shows all of the data in the literature for residual trapping with supercritical CO$_2$ with the coloring distinguishing between sandstone and carbonate rocks. The solid lines represent Land model curves with approximate 95% bounds on the data

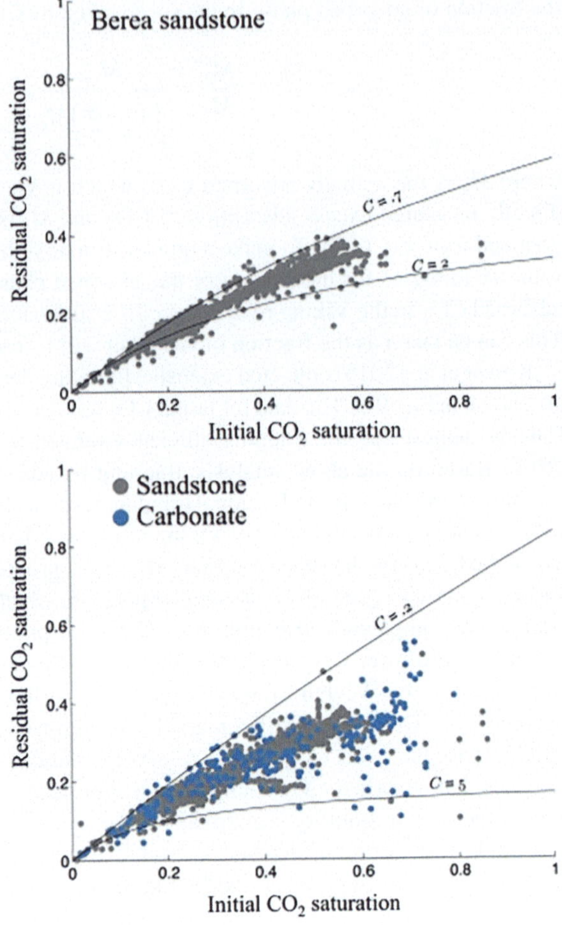


view is valid. However, residual trapping still takes too long to be functional and its contribution to storage capacity is almost negligible in CO$_2$ injection phase (Bachu et al. 2007). When the injection phase is long, residual trapping will start late, but it is in the injection phase that the risk of leakage rises to maximum due to pressure buildup (see, e.g., Espie and Woods 2014). Compared to residual trapping, adsorption trapping occurs instantly from the beginning of CO$_2$ injection and remains in the post-injection phase. In consequence, adsorption trapping decreases overpressure right from the beginning, thereby playing a role in reducing pressure buildup when the risk of leakage is rising. Besides, adsorption can reduce the buoyancy of CO$_2$ plume from the beginning of CO$_2$ injection whereas residual trapping cannot. Another point is that residual trapping only traps discrete CO$_2$ droplets at pore throats so the droplets only occupy a portion of the pore space invaded by reservoir water. By contrast, adsorption occurs at the whole pore surface on the path of CO$_2$ plume with a phase density higher than that of the residual $CO_2$ droplets (1030 kg/m$^3$ vs. 630 kg/m$^3$ at 40 °C and 10 MPa). Therefore, adsorption can give higher trapping efficiency.

## 9.3  Possibility of Bubble Formation

Whereas direct desorption of adsorbed phase into free-phase $CO_2$ would be supressed by the invading water, there is a possibility for $CO_2$ to form bubbles if the water is oversaturated with $CO_2$. The bubbles would tend to move up due to their lower density. However, small bubbles have larger specific surface area and higher Laplace pressure. These conditions should facilitate dissolution of the bubbles. The driving force for the dissolution increases with decreasing size of the bubbles. If the $CO_2$ bubbles coalesce to form larger ones, they will be stopped by pore necks. To go through the pore necks the bubbles must deform and thus increase their surface areas, which requires energy. When the buoyancy force is not enough to push the $CO_2$ bubbles through the necks, the bubbles would be trapped or re-adsorbed. Thus, $CO_2$ bubbles would have much lower mobility compared to the continuous fluid phase. Even if some bubbles squeeze through the pore throats, their rising movements would be hindered by repeated trapping, re-adsorption and increased likelihood of dissolution.

The tendency for the small $CO_2$ bubbles to dissolve may be seen from the Kelvin equation, which describes how the vapor pressure over a curved surface (such as a liquid droplet or gas bubble) differs from the vapor pressure over a flat surface. It is particularly useful in understanding phenomena like nucleation, condensation, and evaporation. The equation is:

$$\ln \frac{p}{p_0} = \frac{2\sigma V_{\mathrm{m}}}{rRT},\tag{9.2}$$

where $p$ is the vapor pressure over the curved interface. $p_0$ is the vapor pressure over a flat interface. $\sigma$ is the interfacial tension between the liquid and the vapor phases. $V_{\mathrm{m}}$ is the molar volume of the liquid. $r$ is the radius of the curved interface, which is positive for liquid droplets and negative for vapor bubbles. The Kelvin equation may also be applied to bubbles as the principles are similar. For gas bubbles in a liquid, the surface tension creates a pressure difference across the bubble interface as given by the Young–Laplace equation:

$$\Delta p = \sigma \left( \frac{1}{r_1} + \frac{1}{r_2} \right)\tag{9.1}$$

For spherical bubbles the equation becomes

$$\Delta p = \frac{2\sigma}{r}\tag{9.1'}$$

The stability of gas bubbles in a liquid is influenced by the interfacial tension, similar to how interfacial tension between liquids and their vapors affects droplets. Smaller bubbles have higher internal pressure due to the interfacial tension, which can affect their growth, dissolution, or stability. The gas bubbles tend to minimize the interfacial area. However, this will increase the Laplace pressure and cause the bubbles to redissolve. Accordingly, only when the bubbles are above a certain critical size can they avoid the redissolution. Meanwhile, the molecules of the dissolved tiny bubbles can diffuse upwards under a concentration gradient and join the large bubbles. Such a concentration gradient exists in reservoirs because the reservoir pressure increases with depth which increases the driving force for gases to dissolve. This process is known as Ostwald ripening, which can lead to the migration of gas molecules upward (Blunt 2022). The diffusion process is governed by the Fick's law:

$$J = -D\frac{dC}{dz}, \qquad (9.3)$$

where $J$ is the diffusion flux; $D$ is the diffusion coefficient; $C$ is the concentration; $z$ is the distance on the direction of diffusion. Because of the diffusion, $CO_2$ bubbles retained in pore space by residual trapping would not stay indefinitely but move upwards eventually. However, this process is expected to be very slow in real reservoirs and could take millions of years for the bubbles to reach the top seal (Blunt 2022). In the case of adsorbed $CO_2$ phase, oversaturation of local water and lack of convective mixing could lead to Ostwald ripening. However, the process would be very slow as well due to slow diffusion of $CO_2$ in water. In principle, diffusion could also occur in the adsorbed phase. However, the rate would be governed by the Fick's law. Unlike the gas bubbles which can have an upward concentration gradient in water, the density of the adsorbed phase can be considered as independent of pressure according to Polanyi's theory. Consequently, there is no density difference within the adsorbed phase and hence no driving force for upward diffusion. Moreover, as has been discussed earlier, the adsorbed $CO_2$ phase would not stay as a continuous phase but dissolve in water in larger pores. As a result, the undesorbed $CO_2$ would become disconnected islands in narrower pore spaces and the Ostwald ripening effect for the adsorbed phase is unlikely to be stronger compared to $CO_2$ bubbles.

## References

A.S. Al-Menhali, H.P. Menke, M.J. Blunt, S.C. Krevor, Pore scale observations of trapped $CO_2$ in mixed-wet carbonate rock: applications to storage in oil fields. Environ. Sci. Technol. **50**(18), 10282–10290 (2016). https://doi.org/10.1021/acs.est.6b03111

A. Baban, A. Keshavarz, R. Amin, S. Iglauer, Impact of wettability alteration on $CO_2$ residual trapping in oil-wet sandstone at reservoir conditions using nuclear magnetic resonance. Energy Fuels **36**(22), 13722–13731 (2022). https://doi.org/10.1021/acs.energyfuels.2c02933

**Fig. 10.2** Conceptual representation of the time evolution of the areal extent of $CO_2$ plume and the associated pressure buildup (from Bachu 2015)

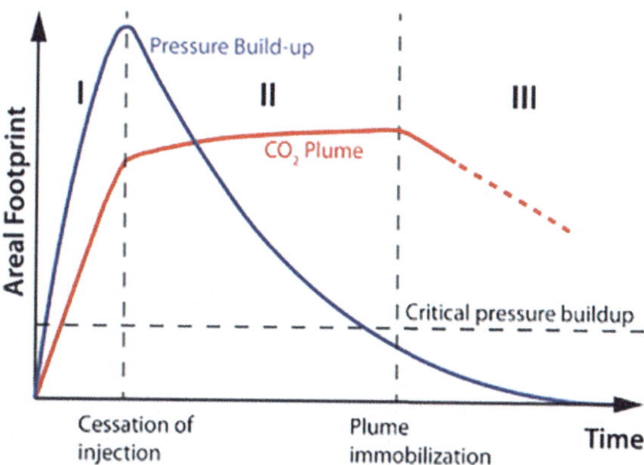


# References

S. Bachu, Review of $CO_2$ storage efficiency in deep saline aquifers. Int. J. Greenhouse Gas Control **40**, 188–202 (2015). https://doi.org/10.1016/j.ijggc.2015.01.007

T. Espie, A. Woods, Testing some common concepts in $CO_2$ storage. Energy Proc **63**, 5450–5460 (2014). https://doi.org/10.1016/j.egypro.2014.11.576

S. Krevor, M.J. Blunt, S.M. Benson, C. Pentland, C. Reynolds, A. Al-Menhali, B. Niu, Capillary trapping for geologic carbon dioxide storage—from pore scale physics to field scale implications. Int. J. Greenhouse Gas Control **40**, 221–237 (2015). https://doi.org/10.1016/j.ijggc.2015.04.006

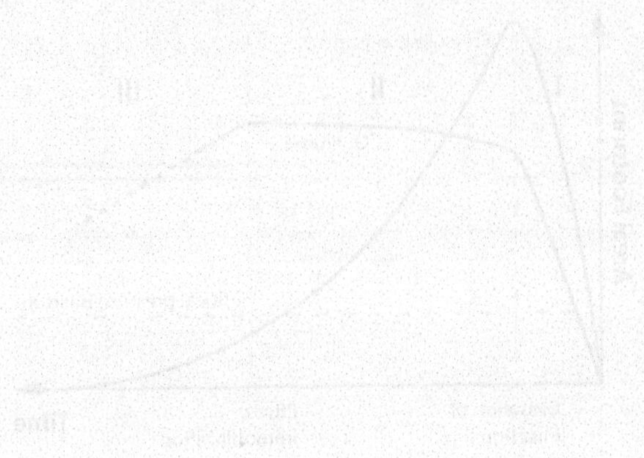

# Chapter 11
# Impact of Adsorption on Security and Safety of $CO_2$ Storage

The security and safety of $CO_2$ storage can be impaired by overpressure, mobility in the post injection phase, and the buoyancy of $CO_2$. If the strengths of the three factors are reduced, security and safety will be enhanced.

Reducing the strengths of overpressure, mobility, and buoyancy can significantly enhance $CO_2$ storage in several ways. Lowering overpressure can help maintain the structural integrity of geological formations. Excessive pressure can lead to fractures or faults, which may compromise containment. By reducing overpressure, the risk of leakage is minimized, ensuring that $CO_2$ remains securely stored. Decreasing the mobility of $CO_2$ within the storage reservoir can enhance containment. When $CO_2$ is less mobile, it is less likely to migrate through rock formations or escape into surrounding environments. There are investigations on techniques such as increasing the viscosity of $CO_2$ or utilizing trapping mechanisms (like capillary trapping) to reduce the mobility. Reducing buoyancy can decrease the tendency of $CO_2$ to rise toward the surface. In geological formations, the lower density of $CO_2$ fluid causes it to migrate upward. By managing conditions to reduce buoyancy—such as through the injection of $CO_2$ at greater depths, storage effectiveness can be improved. As discussed in previous chapters, $CO_2$ adsorption can lower the overpressure, the mobility and the buoyancy, thereby enhancing the security and safety for $CO_2$ storage. However, in addition to the above factors, the security and safely relies largely on the sealing capability of caprocks. In the next section we discuss the potential impact of $CO_2$ adsorption on caprocks.

## 11.1 Effect of $CO_2$ Adsorption on Caprock

Caprock sealing is a critical component in $CO_2$ storage, as it prevents $CO_2$ from migrating to the surface or other permeable layers. We have yet to consider the impact of $CO_2$ adsorption on the sealing ability if $CO_2$ enters caprocks. Leakage of

© The Author(s), under exclusive license to Springer Nature Switzerland AG 2025
J. Wang et al., *Carbon Dioxide Adsorption in Rock and Geological Storage of Carbon*,
SpringerBriefs in Applied Sciences and Technology,
https://doi.org/10.1007/978-3-031-90218-5_11

CO$_2$ through caprock can occur in three ways (see, e.g., Ali et al. 2023): (1) Rapid leakage resulting from mechanical failures of caprocks; (2) Long-term CO$_2$ leakage due to capillary breakthrough; and (3) CO$_2$ diffusion into pore water of caprocks. Regarding mechanical failures, adsorption of CO$_2$ on reservoir rocks can reduce the upthrust on caprocks and hence decrease the risk.

CO$_2$ may also be adsorbed in caprocks. Caprocks have pore channels, albeit narrow, filled with water. The pore surfaces of the caprocks would be water wet originally and the capillary entre pressure is counted on to prevent CO$_2$ from entering the pores. CO$_2$ may dissolve into porewater and move through the pore channels by diffusion, but the rate would be very slow. A common concern, however, is that CO$_2$ under high pressure would overcome the capillary pressure and flow through the pore channels. In such cases the sealing capacity of the caprock is considered to be lost. The high pressure which overcomes the capillary pressure could be caused by high columns of CO$_2$ that form underneath the caprock. Because of the buoyancy forces of the CO$_2$ phase, which increase with increasing column height, the pressures that the caprocks must withstand at the tops of the columns are higher than the pressures in the surroundings. The local high pressure can allow CO$_2$ to displace the pore water and penetrate the caprock. For this reason there is a limit to the column height of CO$_2$ for the caprock to resist the buoyant forces of CO$_2$ effectively. Another concern is that CO$_2$ could de-wet the pore water in caprocks and cause the pore channels to be open to CO$_2$ flow. De-wetting of water phase could also lead to reduced capillary sealing in areas with existing fractures or capillary breaks, potentially creating pathways for CO$_2$ leakage. Accordingly, the capillary sealing capacity relies on the presence of pore water, which provides a capillary barrier to resist the upward movement of CO$_2$. However, if the rock becomes less water wet, CO$_2$ would migrate more easily through the caprock, thereby reducing its sealing effectiveness. As CO$_2$ adsorption could reduce the water wetness of rock pore surfaces or cause de-wetting of the pore water, the capillary pressure that opposing CO$_2$ penetration could decrease and thus the seal's effectiveness could decrease.

There is a view that caprocks have very low permeability and will not allow significant CO$_2$ flow even if CO$_2$ breaks through the capillary barrier (Mirzaei-Paiaman and Okuno 2024). Low permeability is a prerequisite for caprocks. The low permeability implies small pore size (see, e.g., Iglauer et al. 2015a, b) and poor pore connectivity. As has been discussed earlier, small pore size is favorable for CO$_2$ adsorption. The adsorbed CO$_2$ could lead to further decrease of the permeability via pore filling and, in some cases, clay swelling.

The overall effectiveness of the caprock as a seal would depend on the interplay between capillary pressure, buoyancy, and the wettability change. Whereas CO$_2$ adsorption could increase CO$_2$ wetness and enhance the retention of CO$_2$ within reservoirs, it may also reduce the sealing capability of the caprocks. While such situations are unfavorable, CO$_2$ adsorption is a fact which occurs regardless of whether we want it or not. Here we consider several effects of CO$_2$ adsorption which could contribute to reducing the risk of CO$_2$ leakage by sealing the leak pathways.

CO$_2$ may cause swelling of rock minerals such as clays and plugging of voids by adsorbed CO$_2$. Most caprocks, including shale, contain clays as dominant mineral species. Clays can have significant capacities to adsorb $CO_2$. Swelling of some clays such as smectites can occur due to adsorption of $CO_2$ (Busch et al. 2016). The swelling may create a denser, more cohesive matrix in the caprock, potentially increasing the structural integrity and resistance to fracturing. It may also result in closure of microfractures and filling pore spaces which can increase the overall impermeability of caprocks and therefore decrease the risk of $CO_2$ leakage. However, excessive swelling can lead to mechanical instability in the caprock by creating internal stresses that may cause cracking or fracture propagation, thereby compromising the seal. Accordingly, whereas the swelling of clay minerals due to adsorption of $CO_2$ can potentially enhance the sealing effectiveness of caprocks, it may also introduce mechanical challenges. The overall impact of the swelling on caprock integrity depends on the extent of swelling, the specific mineral composition of the caprock, and the geomechanical context of the storage site. Proper assessment of the swelling effect is vital to storage integrity.

The adsorption of $CO_2$ in caprocks predominantly occurs on the inorganic clay minerals and, in the case of shales, on organic matter (Iglauer 2018, 2017; Iglauer et al. 2015a, b; Pan et al. 2018; Busch et al. 2008). Busch et al. (2008) postulated that the inorganic content of shale caprocks is of primary influence on its adsorption capacity. Experimental results on a caprock shale with a total organic content (TOC) of < 0.5% suggested that the high adsorption capacity could be entirely attributed to the inorganic clay content. Clay minerals have high capacity for adsorption of $CO_2$ and have shown to be effective in capturing and storing $CO_2$ (Raveendran et al. 2024).

## 11.2  The Potential of Pore Filling by Adsorbed $CO_2$

Caprocks are characterized by small pore size in the range of micropores, which result in higher adsorption capacity compared to reservoir rocks. When buoyant $CO_2$ enters the micropores, the pressure will be high. As a result, pore filling by adsorbed $CO_2$ phase could occur. Considering the natural pressure gradient in the caprock, even at the upper side of the caprock the pressure could still be high enough to keep $CO_2$ in the adsorbed phase. Besides, the adsorbed phase in the micropores is attached to the pore walls and not buoyant. Thus, the bulk of the adsorbed phase in the micropores will be immobilized. Although it is possible for adsorbed molecules to diffuse from high pressure side to the low pressure side, the process will be slow. As has been discussed earlier, the density of adsorbed $CO_2$ may be independent of pressure according to the Polanyi theorem. Therefore, there is no density gradient which provides the driving force for $CO_2$ diffusion in the adsorbed phase, so significant upward migration of $CO_2$ through diffusion in the caprock is not likely. Moreover, adsorbed $CO_2$ may occupy the total pore surface, as seen from the adsorption data of a top seal shale studied by Busch et al.(2008). In that study 1 kg shale could adsorb 1 mol $CO_2$ (Fig. 11.1), suggesting that the adsorbed $CO_2$ phase would occupy the whole pore space and block the migration pathway of $CO_2$. Here we give a quick evaluation.

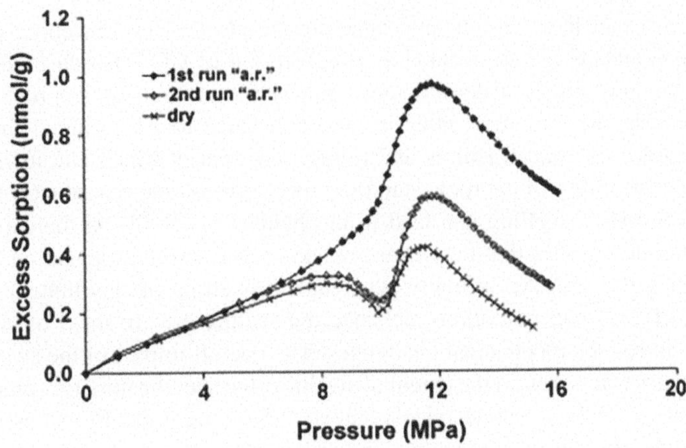


**Fig. 11.1** Excess adsorption data for CO$_2$ on a top seal shale studied by Busch et al. (2008)

The porosity of the shale was reported to be 20%. The grain density of the shale was 2.7 g/cm$^3$. The temperature of the adsorption measurement was 45 °C, and a maximum value of excess adsorption 1 mmol/g was reached at 12 MPa. Under this temperature and pressure condition the CO$_2$ fluid density would be 658 kg/m$^3$. Using the formula of effective density in Chap. 4,

$$\rho_{\text{eff}} = \rho_f\left(1 + \frac{\rho_r}{\rho_f} \frac{(1 - \varepsilon)}{\varepsilon} \frac{M_{\text{ex}}}{M_r}\right) \tag{4.8}$$

the corresponding density of CO$_2$ in the pore space of the shale can be calculated to be 1133 kg/m$^3$. This value is higher than the density of adsorbed CO$_2$ 1030 kg/m$^3$ obtained from the van der Waals constant $b$, which suggests that the pore space of the shale would be fully filled by adsorbed CO$_2$. This can be seen with the aid of the formula for the volume-weighted average density of CO$_2$ in pore space

$$\rho_{\text{av}} = \frac{\rho_a V_a + \rho_f V_f}{V_a + V_f} = \frac{\rho_a V_a + \rho_f V_f}{V_P} \tag{4.11}$$

which has been shown in Chap. 4 to be equivalent to the effective density. Based on Eqs. (4.8) and (4.11) the following relation can be obtained:

$$\rho_a = V_f(\rho_{\text{eff}} - \rho_f)/V_a + \rho_{\text{eff}} \tag{11.1}$$

Because $\rho_{\text{eff}} > \rho_f$, $\rho_a$ must be higher than $\rho_{\text{eff}}$ if $V_f > 0$. In other words, if the pore space was only partly occupied by adsorbed CO$_2$, the density of the adsorbed CO$_2$ would be even higher. On the other hand, if we use the van der Waals value 1030 kg/m$^3$ for the density of adsorbed CO$_2$ to calculate the true adsorption using the relation

$$M_a = M_{ex} / \left( 1 - \frac{\rho_f}{\rho_a} \right) \qquad (3.10')$$

we will get a value 10.3 wt% of adsorbed $CO_2$ per unit mass of the rock, which can fill 10% more than the pore space of the 20% porosity rock. Accordingly, it is reasonable to expect that the pore space of the shale can be fully occupied by adsorbed $CO_2$ through pore filling, leaving no room for free-phase $CO_2$. High density of the adsorbed phase can yield stronger bonding, as more adsorbed molecules interact with the solid surface. The resultant pore plugging should enhance the overall sealing strength and make caprocks more impassible to fluid phases.

The above result is consistent to the accepted pore-filling mechanism for $CO_2$ adsorption in shales. Pore filling occurs when fluid molecules are adsorbed into the available surface area within the micropores, filling up the available space. In micropores, the surface area per unit volume available for adsorption is very large, and the proximity of the pore surfaces enhances the interaction between fluid molecules and the surface, allowing a higher density of adsorption compared to larger pores or macropores. It is worth noting that adsorption capacity of shales is usually attributed to their organic contents. However, in the above-discussed shale the organic content was below 0.5% so the primary adsorption capacity was considered to be from the clay content. Busch et al. (2016) further postulated that clay content dominates the adsorption capacity of caprocks. Based on this view and the pore-filling mechanism, it can be envisaged that, once entering the micropores, $CO_2$ molecules can be readily trapped as they approach the pore surface. A barrier could form soon as the pores are filled by adsorbed $CO_2$. The barrier could prevent further advancement of $CO_2$ in the caprock. A consequence of this situation is that $CO_2$ plugs are attached to the pores and no longer buoyant. For adsorbed $CO_2$ to penetrate further, desorption and dissolution are required. Desorption would be limited by the slow rate (as discussed in Chap. 9), and by the pressing from pore water. Dissolution will be limited by mass transport of dissolved $CO_2$ molecules from saturated layers to low concentration zones. Unlike the situation inside reservoirs, where $CO_2$ saturated water can sink deeper due to its higher density, within the caprock the sinking-enabled mass transport will not happen because of the plugged pores. Without natural convection forces the $CO_2$ molecules will need to depend on diffusion only to move away from the saturated zone. In consequence, the driving force for $CO_2$ penetration in caprocks will be low. For the $CO_2$ to escape through the caprock layer, the process will not be faster than the scenario evaluated by Busche et al. (2016), which may take geological time scale.

Over the long term, the impact of $CO_2$ on caprock wettability and sealing properties may also depend on other interactions that occur. For example, mineral reactions might create secondary barriers to enhance sealing capacity, in spite of changes in wettability.

After all, $CO_2$ leakage through the tiny pores in caprocks due to decreased capillary trapping can be an important issue but a minor concern compared to other leakage paths including faults and abandoned wells. Even if wettability permits the $CO_2$ to push through, extremely low permeability would lead to long time scales (thousands of years, for example, depending on caprock thickness) before $CO_2$ breakthrough and therefore pose little risks in short to medium terms (Iglauer et al. 2015a, b; Busch et al. 2016).

In summary, the sealing capacity of caprocks may be lower than expected. Adsorption of $CO_2$ on caprocks further complicates the situation. On the other hand, adsorbed $CO_2$ could make the rocks more impenetrable. According to the adsorption data, the pore space in caprocks will be filled with adsorbed $CO_2$, even when the porosity is large. However, there are a lot of unknowns and uncertainties. Further research is warranted to ensure the effectiveness of caprock sealing and $CO_2$ storage security.

# References


F. Ali, B.M. Negash, S. Ridha, H. Abdulelah, A review on the interfacial properties of caprock/$CO_2$/brine system-implications for structural integrity of deep saline aquifers during geological carbon storage. Earth-Science Rev. **247**, 104600 (2023). https://doi.org/10.1016/j.earscirev.2023.104600

A. Busch, S. Alles, Y. Gensterblum, D. Prinz, D.N. Dewhurst, M.D. Raven, H. Stanjek, B.M. Krooss, Carbon dioxide storage potential of shales. Int. J. Greenhouse Gas Control **2**(3), 297–308 (2008). https://doi.org/10.1016/j.ijggc.2008.03.003

A. Busch, P. Bertier, Y. Gensterblum, G. Rother, C.J. Speirs, M. Zhang, H.M. Wentinck, On sorption and swelling of $CO_2$ in clays. Geomech. Geophys. Geo-Energy Geo-Resour. **2**(2), 111–130 (2016). https://doi.org/10.1007/s40948-016-0024-4

S. Iglauer, C.H. Pentland, A. Busch, $CO_2$ wettability of seal and reservoir rocks and the implications for carbon geo-sequestration. Water Resour. Res. **51**(1), 729–774 (2015a). https://doi.org/10.1002/2014WR015553

S. Iglauer, A.Z. Al-Yaseri, R. Rezaee, M. Lebedev, $CO_2$ wettability of caprocks: implications for structural storage capacity and containment security. Water Resour. Res. **51**(1), 729–774 (2015b). https://doi.org/10.1002/2015GL065787

S. Iglauer, $CO_2$–water–rock wettability: variability, influencing factors, and implications for $CO_2$ geostorage. Acc. Chem. Res. **50**(5), 1134–1142 (2017). https://doi.org/10.1021/acs.accounts.6b00602

Iglauer, Optimum storage depths for structural $CO_2$ trapping. Int. J. Greenhouse Gas Control **77**, 82–87 (2018). https://doi.org/10.1016/j.ijggc.2018.07.009

A. Mirzaei-Paiaman, R. Okuno, Critical review on wettability, optimal wettability, and artificial wettability alteration in rock-brine-$CO_2$ systems for geologic carbon sequestration. Gas Sci. Eng. **132**, 205499 (2024). https://doi.org/10.1016/j.jgsce.2024.205499

B. Pan, Y. Li, H. Wang, F. Jones, S. Iglauer, $CO_2$ and $CH_4$ wettabilities of organic-rich shale. Energy Fuels **32**, 1914–1922 (2018). https://doi.org/10.1021/acs.energyfuels.7b01147

G. Raveendran, K. War, D.N. Arnepalli, V.B. Maji, Modelling carbon dioxide adsorption behaviour on montmorillonite at supercritical temperatures. Adsorption **30**, 1703–1716 (2024). https://doi.org/10.1007/s10450-024-00525-z

# Chapter 12
# Conclusions and Recommendations

Adsorption results in a phase change. The adsorbed phase has higher density, low mobility and significant affinity to the pore surface. The higher density enables higher storage capacity and lower overpressure, effectively increasing the pressure space of the reservoirs. The immobilization by adsorption results in a smaller volume of mobile $CO_2$, decreasing stress to the caprock and increasing storage security. The affinity to the pore surface can change the wettability and decrease the capillary pressure. This should increase $CO_2$ injectivity and sweep efficiency. Moreover, more adsorptions can take place due to the decreased capillary pressure in narrow pores, increasing utilization efficiency of pore space. In addition, $CO_2$ desorption can be very slow, allowing keeping adsorbed $CO_2$ immobilized and increasing $CO_2$ dissolution in water, thereby contributing to solubility trapping and mineral trapping. According to the analysis in this work, adsorption can immobilize the $CO_2$ plume eventually in the post-injection phase, similar to the immobilization effect expected for residual trapping but in a more efficient way by resulting volume reduction of $CO_2$.

The impact of adsorption on caprocks needs further investigation because of the possibilities of decreasing the water-wetness of pore surface and dewetting of water which reduce the strength of capillary sealing. On the other hand, decreased water-wetness implies increased $CO_2$ adsorption, which could lead to filling the micropores of caprocks with denser and immobile adsorbed $CO_2$ phase. Adsorption of $CO_2$ could also result in clay swelling. The pore filling by adsorbed $CO_2$ and pore closure by clay swelling may block leakage pathways. However, excessive stress caused by the swelling may affect the mechanical stability of the seal. There are a lot of uncertainties regarding the interactions of $CO_2$ with the caprocks. Nevertheless, immobilization of $CO_2$ due to adsorption in caprocks should put an end to the buoyancy effect and hence stop the tendency of moving upward. Moreover, pore filling by the immobilized $CO_2$ should further decrease, if not fully eliminate, the already extremely low permeability of caprocks.



J. Wang et al., *Carbon Dioxide Adsorption in Rock and Geological Storage of Carbon*,
SpringerBriefs in Applied Sciences and Technology,
https://doi.org/10.1007/978-3-031-90218-5_12

As $CO_2$ adsorption can result in great differences to $CO_2$ storage in multiple ways, it should be included in modeling and simulation for $CO_2$ storage, regarding:

1. Capacity Estimation: Adsorption affects the effective storage capacity of the reservoir. Models need to account for the amount of $CO_2$ that can be adsorbed onto the pore surfaces in addition to the volume of $CO_2$ that can be held in the pore spaces. This will increase the total amount of $CO_2$ that can be stored.
2. Migration Dynamics: Adsorption alters the mobility of $CO_2$ within the reservoir. As adsorbed $CO_2$ would be immobilized and hence decrease the buoyancy of $CO_2$ mass, $CO_2$ plume spreading and pressure propagation would be largely impacted. As a result, simulated $CO_2$ behaviors would be quite different.
3. Long-term Storage Stability: Understanding adsorption dynamics can help assess the long-term stability of $CO_2$ storage. When $CO_2$ is adsorbed in reservoir rocks, it would be much less likely to migrate and leak, which can enhance the overall security and safety of the storage.
4. Geochemical Reactions: Adsorption can influence geochemical reactions between $CO_2$, and the mineral surfaces, affecting rock properties and potentially leading to mineralization over time. Models need to consider these interactions to predict the long-term consequences of the reactions.

Incorporating these factors into models and simulations can lead to more accurate predictions of $CO_2$ behaviors in storage reservoirs, ultimately improving the effectiveness and safety of $CO_2$ sequestration efforts.

$CO_2$ adsorption can also impact monitoring of $CO_2$ storage projects. Long-term monitoring is indispensable for $CO_2$ storage to make sure that the $CO_2$ remains securely stored. For this purpose, the movement and growth of the $CO_2$ plume should be tracked. Effective monitoring requires understanding the behavior of $CO_2$ in the subsurface and its interactions with reservoir water and rocks to avoid any possible leakage and contamination of ground water. Predictive models can guide real-time monitoring strategies. By understanding the expected movement and pressure changes caused by injected $CO_2$ operators can better design monitoring systems to detect anomalies like unexpected $CO_2$ migration or pressure buildup. Understanding how $CO_2$ will behave in the subsurface is also important to data analyses and interpretation, enabling better forecasts and real-time updates. As has been discussed, adsorption of $CO_2$ can greatly affect the migration behavior and pressure buildup. Taking into account the adsorption can improve the efficacy of $CO_2$ monitoring significantly.

Overall, adsorption of $CO_2$ in rocks can impact $CO_2$ storage profoundly. It should become a routine part in reservoir assessment for CCS projects. The authors hope that the results and methods presented in this work will support efficient use of available geological resources for $CO_2$ storage as part of the efforts for climate change mitigation and the net-zero transition.

The manufacturer's authorised representative in the EU is Springer
Nature Customer Service Centre GmbH, Europaplatz 3, 69115 Heidelberg,
Germany. If you have any concerns regarding our products, please
contact ProductSafety@springernature.com

Printed and bound by CPI Group (UK) Ltd, Croydon, CR0 4YY
01/07/2026
02153169-0001